Wild Goose Marsh
Horicon Stopover

Wild Goose Marsh
Horicon Stopover

by Robert E. Gard

photography by Edgar G. Mueller

WISCONSIN HOUSE, LTD.
MADISON, WISCONSIN

First Edition
Library of Congress Card Number 72-89568

Printed in the United States of America for Wisconsin House, Ltd.,
by Straus Printing and Publishing Co., Inc., Madison, Wisconsin

For Jane Inge
and
All the other great people who have
given their lives to the preservation
and training of human talent and
to the conservation of our
natural resources.

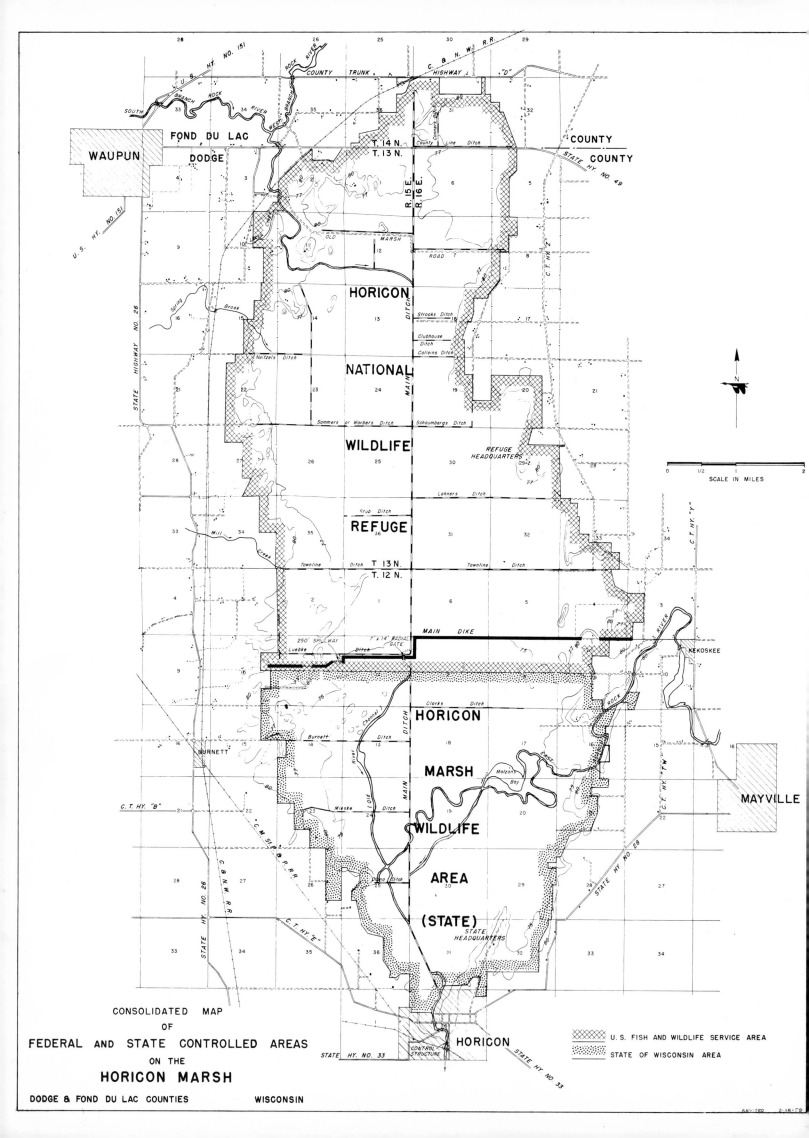

CONSOLIDATED MAP
OF
FEDERAL AND STATE CONTROLLED AREAS
ON THE
HORICON MARSH

DODGE & FOND DU LAC COUNTIES   WISCONSIN

⧖⧖⧖ U.S. FISH AND WILDLIFE SERVICE AREA

∴∴∴ STATE OF WISCONSIN AREA

TABLE OF CONTENTS

HORICON MARSH

Horicon Marsh, an area of 31,653 acres, was scoured out by the Wisconsin glacier, at least 10,000 years ago. Gradually the upper Rock River made deposits which slowed its current and spread its waters over the marshland. The Marsh became a haunt of the earliest Indians whose mounds remain. To promote lumbering, transportation, and agriculture white pioneers built a dam in 1846. Horicon Lake, covering 51 square miles, became famous for hunting and fishing. The dam was removed in 1869, restoring the Marsh, which was subjected to various development schemes that changed its character. Climaxing a twenty year struggle by conservationists, Horicon National Wild-Life Refuge was established July 16, 1941. The State controls the south 10,857 acres; the Federal government, the north 20,796. A wide range of wild fowl, many varieties of small birds, and numerous fur-bearing animals constitute the population of Horicon Marsh.

Erected 1959

ACKNOWLEDGMENTS

I could not have assembled this book without the memories, concern and willingness of many people. It grew out of travels and conversations. Ed Lechner, Pete Feucht, Eddie Lehner, Bob Personius, Wilton Erdman and Barney Wanie were all of invaluable assistance. The personnel of the Marsh's Federal and State Headquarters generously extended their time and interest. I am also indebted to Martha Krueger, Harold Mathiak, Bill Schreiber, Pat Burhans, August Krueger and John Strook. Interviews with Butch Burkhardt, Hubert Pagel, Adolph Kruel, Clyde Huffman and Dr. Fred Karsten as well as Mr. Franklin Koch of the Horicon area added a great deal.

Special thanks go to Edgar G. Mueller who in picture and word contributed so much.

All of these people and more, Wes Burmeister and William Steuber, and L. G. Sorden helped me. Marian and Mark Lefebvre and, of course, my secretary, Arnita Ready, were invariably on board to assist. I owe them more than I can say.

In addition, I am indebted to others who have worked with the Marsh and whose writings have served as source and stimulus: B. H. Tallmadge, *Horicon Marsh — Its History And Development* (1914), Joseph Schafer, *The Winnebago-Horicon Basin; A Type Study In Western History* (1937), Allie E. Freeman and Walter R. Bussewitz, *History Of Horicon* (1948), and Giles F. Clark, *Horicon Marsh* (1969). *The Horicon Reporter* and *The Wisconsin Magazine Of History* clarified and aided my research. *The History Of Dodge County* and proceedings of The State Historical Society of Wisconsin, particularly Vol. XII containing Thwaite's account of the Black Hawk War, broadened my knowledge. Other source materials: Royal Brunson Way, *The Rock River Valley* (1926), A Federal Writers' Project, *Wisconsin Indian Place Legends* (1936), Harry Ellsworth Cole, *Stagecoach And Tavern Tales Of The Old Northwest* (1930), and James M. Phalen, *Sinnissippi* (1942). Of singular help was Milo M. Quaife, *Published Writings Of Milo M. Quaife*, 1910-1955 (1956).

I have examined that strange pamphlet *Onions and Independence* published by the drainage developers of the Marsh, and I have depended on the *Milwaukee Journal* for assorted materials. A brief history of the Marsh shooting club by Walter Frautschi was helpful; so was Virginia Palmer's excellent account of the Marsh in the *Wisconsin Magazine of History*. From the old files of *The Wisconsin Archeologist* I depended for Dr. Bruder's account of Marsh explorations. There are doubtless scores of other items that should be mentioned, such as the Canadian Government folder on the Canada goose; and the public distribution materials prepared by the Wisconsin Department of Natural Resources. Walter Scott and Frank King of that staff were of particular assistance. I particularly wish to thank Jim Bell of the State Marsh Headquarters who allowed his staff to assist me, and let me examine many rare materials in his files. I am also indebted to my own book, *Johnny Chinook,* Longmans, Green and Company, for the story of the great goose capture on St. James Bay.

Thousands Of Wings Take To The Air — Mueller's World-Famed Picture

PHOTOGRAPHER'S FOREWORD

The Horicon Marsh has directly and/or indirectly affected the lives of many. To us the Marsh has always been a focal point for pictures and news stories. Controversial as the Marsh has been, there were times when we, too, had mixed feelings about restoration back in those "fighting days" of the late thirties and early forties. Restoration backers were thinking of wildlife promotion in terms of ducks and fishing. The price of course meant flooding farmers' fields, much of it good crop land. Our sympathies included those farmers who had to leave their homes and start over someplace else.

All this is now history. And that is what this book is all about.

Man has contributed much over the years developing the Marsh.

But the wild goose has made it what it is today. Never in their fondest dreams did the Marsh planners of thirty years ago realize what that great bird, the Canada goose, would make Horicon Marsh today!

By its very nature, shape, and form the wild goose is inspirational. Its migrations and flight patterns have pointed the way and seasons for man through the march of time. Its shape and form have been followed by modern designers in the production of countless vehicles of motion, culminating in the giant man-made birds of the skies that hourly carry thousands from one part of the world to another, and even into outer space, and the moon.

When Bob Gard asked us to become a member of his team in the production of this book it meant retracing our steps.

Going back through the years via thousands of negatives and photographs showing highlights of Horicon Marsh activities, wildlife and development progress, we relived the countless treks we made to the Marsh, sometimes climbing windmill towers and silos and always carrying our Speed Graphic and Graflex cameras (they were heavier and larger than the compact 35-millimeter cameras we use today) for a better view. Or racing a fast-moving grass fire closeup for a "hot" picture.

From this lifetime collection of pictures the editors selected the views you see reproduced in this book. Picture editors are human beings. They are not infallible. And, because the selection of pictures is entirely a matter of judgment (what you and I may like may not be someone else's cup of tea) we can only hope that the reader may agree with the selection.

With respect to one of the pictures you will find perhaps a good reason for the editors' unanimous choice, as a jacket picture. This photograph, made October 23, 1964, has been reproduced in so many publications the world over that we have frequently been asked how it was made.

On a sunshiny Sunday morning, while crossing Highway 49, running east from Waupun, Wisconsin, en route to a family outing, we noted an unusually large concentration of birds on the refuge. We stopped so our friends might view the spectacle. Rather than detain our friends by remaining on the scene to take pictures we decided to return alone the next day.

Monday morning's weather was not the bright and sunshiny yesterday. It was overcast and there was some fog. But the geese were there, concentrated like I had never seen them before, or since. Working in a routine fashion, we shot picture after picture.

The best was yet to come.

After an hour, or perhaps two (I seem to lose all track of time while bird-watching through a camera lens), we thought we had enough shots and were about to pack up and leave, when suddenly, at infinity (perhaps 300 yards in the distance) thousands of wings took to the air alerting the birds in the foreground — heads erect, bodies braced for takeoff — soon as there was flying space overhead. We released the shutter at 1/250th of a second capturing this "once in a lifetime" picture.

The picture prompted Quincy Dadisman, then state editor of the *Milwaukee Sentinel* (first to use it) to comment, "You've got a real winner here!"

Since that first printing this picture has been reproduced many times. Its use and popularity continue to mount, gaining ever more fame for Horicon Marsh via the wings of the wild Canada goose.

Many of the picture opportunities we have had through the years resulted directly from tips and suggestions from people thoughtful enough to call our attention to picture possibilities.

We are grateful.

We salute the wild goose for setting its seasonal flight patterns over our back yard, settling down on the vast Horicon Marsh to a smorgasbord of feed and a long rest for all to see.

But most of all we thank the Good Lord for making it all possible.

Edgar G. Mueller

The Trail Tree Waits: Vine-Tied Branches Pointed The Way

Perhaps this foreword, or opening ramble, isn't really needed. It may actually be redundant, because I wrote the foreword a long while after I had written the opening few pages of the book — and they may reflect each other a little. But even risking redundance, I am including the foreword anyway, because I think it represents an enlargement of a curious and all-but-forgotten happening that inspired the creation of *Wild Goose Marsh — Horicon Stopover*. As a dramatist, I am ever aware anyway of the fascination of cause and effect. It may well have been that this book was actually preordained to be written. I am believer enough in the fates, or the strange way chance sometimes works things out in terms of unbelievable coincidence; or in the old Indian Gods and their timeless wills . . . something, at least, impelled me to produce this book about the Great Horicon Marsh.

The way it all started was when I visited Horicon, Wisconsin, at the invitation of a Kathleen Karsten, away back about 1952. I made some kind of an informal talk that evening, I remember, to a group of local ladies. Then I was taken to dinner at the Rogers Hotel in Beaver Dam. During the pleasant meal I was told by Kathleen about a wondrous tree that stood near Horicon, near the Great Marsh, and that it was a trail marker tree . . . one which the old Indians had tied with vine certain branches to point in the same direction, and so to mark an important Indian trail up and through the Marsh. The story, as she told it, fascinated me, and I expected some day to use it, but I didn't really come back to her yarn for many years. Then, one afternoon in 1971, driving alongside the Marsh on Highway 28, I had an impulse that the time had arrived to tell the greatest wetlands story in America . . . for so I had come to believe. I was amazed to find that very little popular literature had been written about the Marsh, though the lore about it and about its wildlife in oral terms was immense. Also the Marsh had become one of the truly fascinating wildlife shrines of America, especially in that there is now a greater congregation of Canada geese at Horicon in spring and fall than at any other North American refuge. Millions of visitors arrive to view the wonderful birds, and to watch them feed and rest, but nobody knows very much about the Marsh, or why the birds are there, or about the unlikely chain of happenings that created, finally, out of years of struggle over the fate of the land, a refuge for the birds and a real chance for them to discover their mid-America haven.

The strange part of the story comes now. I was visiting in Mayville, Wisconsin, in March, 1972, with my friend Ed Mueller who has done the photographs, and during lunch I started to tell him and his wife Dorothy about that long ago previous visit to the Marsh, and how Kathleen Karsten had told me about the Indian tree. Ed said he knew a fellow named Willard Bartelt in Horicon who ran the city dairy and lunch, and said that Willard Bartelt might know where that old tree was located. If he did, Ed said, we would go out and photograph it . . . for I had remarked that I wanted to begin and end the book with some symbol that might have significance for both the old days and for the new.

Ed called his friend Bartelt who said that *he* didn't know anything about the tree but that if we would call back in a while he might find somebody who did. Meanwhile I called the home of Kathleen Karsten (whose husband I discovered is now the local doctor in Horicon) to find out whether she remembered the incident that had happened twenty years before, and whether the old tree she so beautifully described was still standing. There was no one home at the Karsten's.

Then we called Willard Bartelt again, and he said, wait a minute, and then he said: I have Dr. Karsten here. He is eating lunch here today. He will talk with you. Funny thing was that it was the first time Dr. Karsten had eaten lunch at the dairy for many months. Kathleen just happened (if that's the way you want to look at it) to have gone to Fond du Lac shopping. It turned out that in the intervening twenty years since I had first heard of the tree from Kathleen, Dr. Karsten had purchased the farm on which the old trail tree stood. He was as excited as I when I reminded him of the long forgotten incident, and of the tale told me then in his home. He got his Scout, for the land was mushy still in the early spring, and drove Ed Mueller and me out to the location of the old tree. There it was . . . at least two hundred years old, we thought, a weathered and beaten old oak, some of the branches dead, but the two Indian-vine-tied ones pointing out unmistakably, and together, the ancient roadway of the Indian peoples. And we stood there for a while and marveled at the way chance works things out, or the fates, or the Indian gods who *might* have decided that it was time for Robert Gard to tell the Horicon Marsh story.

Anyway, Ed Mueller did take the picture of the tree while Dr. Karsten and I exchanged pleasant reminiscences, and the Horicon Marsh trail tree became immutably fixed for me in the folklore of the Horicon Marsh. As I said, I believe in preordained coincidences . . . they have happened to me too often, in too many strange ways. I can't ignore them. So I have produced this book about a fascinating and wonderful bog: The Horicon Marsh; and about its people. And I do thank, from my heart, Kathleen Karsten, and Dr. Karsten and all the fine folks of the whole Marsh area who helped me, and were so kind to me when I was preparing the materials. I only wish that the Indians who tied the tree branches to point the trail were around still to tell their tales, but they have vanished, and I suppose, they will never return unless the strange happenings that must occur in the incomprehensible spaces of time begin all over again in the same way, which maybe they will do, and another Ice Age occurs, and new Marsh areas are created, and migrating red-skinned peoples come again to enjoy the luscious plenty that, little by little, grand Nature, a part of the Great Spirit's plan, provides for us all, if we will but wait, take care, listen, enjoy, tolerate, preserve, and have faith.

Spring-fed Creek

Sometime in 1846 a few early arrivals at a place called Hubbard's Rapids at the south end of a vast marsh in east central Wisconsin gathered at a pioneer's cabin to discuss a proper name for the new settlement. The gathering was held in the tamarack log shack of William Larabee who hailed from Lake George, New York.

Some of the folks present thought that Hubbard's Rapids ought to be called Forest City, but William Larabee put up a violent battle for the name "Horicon." Horicon, it appeared, had been the original name of Lake George and meant clear, pure water. Clear pure water seemed to be what they had a plenty of around Hubbard's Rapids. The assembled pioneers, who probably included Mart Rich who was building a dam for a sawmill at the rapids, thereupon decided that "Horicon" should be the final name for their new town.

These early pioneers probably didn't rightly understand in those days the importance that the old Indian peoples had attached to this great marshy area fed and watered by the Rock River. And very likely they did not know the full significance of the trail marker trees that the Indians had shaped to indicate the ancient paths; or to make hiding places to wait passing game. Actually, I suppose, the pioneers cared little and the trees vanished, one by one, through the subsequent years until, in 1971, only a few were left. There is one still standing at the western edge of the Marsh — an oak that has a "blind" branch where a hunter might waylay a passing deer; and on the eastern side of the Marsh another great oak that retains on its Marsh side two curiously bent branches. The branches are those that Indians generations ago forced from their natural attitudes and tied with vines so that they pointed out the trail. In early winter the branches are etched out starkly, and it is easy for me to imagine in them both an urgency and a silent questioning, for the tree's gnarled indication now seems strangely meaningless. The old trail has long, long been forgotten and the Indian peoples who journeyed that way have entirely disappeared. The last Indian family moved from the Marsh area years ago, and no one of Indian blood now remains of the thousands of that almost endless line, much of it shadowy, stretching backward to the ancient Copper People who were probably around and in the Marsh two or three thousand years before Christ was born; to their equally shadowy inheritors, the Red Ocher People, who placed red ocher on their burials; to the first Woodland Peoples who also arrived before Christ and hunt-

ed through the Marshlands in small, wandering, bands; and then those later woodland peoples who buried their dead in effigy mounds made in bird and animal form and who became known centuries later as the Effigy Mound People . . . of immense industry, apparently, for they left mounds as long as 500 feet and even one hollowed out, or scooped-out "intaglio" near Fort Atkinson in Jefferson County.

And the prehistoric ones, some graced by greater skills in crafts than others, were replaced by the Indians of later centuries still, of Siouan and Algonkian stock, and who were thriving in Wisconsin and around the Marsh when the white man appeared: the Winnebago, Menomini, Potawatomi, Illinois, Kickapoo, Chippewa, Fox, Sauk . . . and so on and on until, when the white man arrived, the Winnebago, the Potawatomi, the Sauk, Fox, Illinois, and maybe the Kickapoo were most numerous at the Marsh.

They all regarded the Marsh as a chief hunting and fishing location. The oak branches that monitored the trail for some of these later peoples seem now but empty symbols that appear absurdly lost in time, and seem to be asking: what has happened? What final ruin has the white man accomplished in his dealings with the Indian's paradise, and what does the Great Marsh now signify in its uneasy relationship to man and to the ineluctable ways of nature?

I first heard of a Marshside Indian trace tree a long while ago. About 1952, I think it was, I was visiting one evening some friends in Horicon and they described the tree and its pointing branches to me in such a dramatic way that I developed then a deep curiosity to see it, and to try to identify the trail the branches had been set so permanently to mark. Because of a press of time I didn't see the tree then, however, but I never forgot the story or the marking branches.

Through a strange series of coincidences, though, which I have described in my foreword to this book, I did finally, twenty years later, see the marker tree, and then in 1972 I stood beneath it trying desperately to grasp a sense of Indian moccasins on a worn, earth walkway; and for an instant I experienced a profound and creative knowledge that the darkly pointing and twisted branches did have meaning, even now, and that I must seek for it.

When I returned with that search as one purpose, hidden in a broader one — to tell the story of the Marsh and of the people as they are now — there was gleam ice on the willows. The faint orange-red of the red osier dogwood at the edges of the Marsh shone through the thin film of ice a little, and the red of the dogwood and the slow drift of snowflakes, and the slight sound uttered by a winter wind among low bushes and grasses of the Marsh made me aware that beyond the roads that white men had created along the margin, and within the Marsh itself, nature held and would always hold a magnitude of mystery.

The Marsh, in winter, has a special beauty. In the early fall the wild migrating Canada geese come, of course, and they stay until late November and then they sweep down the flyway to their winter grounds near where the Ohio flows into the Mississippi. There, the geese's refuges are called Horseshoe Lake, Crab Orchard, and Union County; and the Horicon Marsh from whence the geese have departed lies quiet; the constant talking and honking of the Canadas is heard no longer, the morning blast of smokeless powder from the Marsh margins has died away until next season, and the wind begins to comb the tall grasses and to bend the cattails that are now so plentiful. The farmers along the Marsh edge, now that their crops are threatened no longer by the ravenous Canada geese, look ahead to the quiet winter, and to the memories that remain to some older ones, of farms that were once near, or partly in, the Marsh itself, and which had had to be sold to the State of Wisconsin and to the Federal land buyers when the north and south portion of the Marsh were taken over for the wildlife refuge — beginning in 1927, the south part, by the State; and beginning in 1941 — the north part, by the Federal Government.

There is some bitterness among the farmers, for the Marsh, which was home to these sturdy, homeplace people, mostly of German descent on the east, and on the west side, the "Hollanders," stubborn, strong, with opinions undulled by passing time; the Marsh, their home and the flavoring of their lives, drove deeply into their emotions. The present generations love the birds and the grasses and the smell of the Marsh, but they remember how it was, and how their old folks spoke so nostalgically of it, and the sacrifices that many pioneers made simply to live in or near the Marsh and to be a part of its seasonal change and of its always near and teeming wildness.

Most of the old-timers who lived there thought of themselves as part man, part wild bird, part muskrat, beaver, mink, fox, and raccoon for they loved and understood the animals, yes, and hunted them, and still, protected them in their own way; and most of the old-timers lived on the Marsh because they loved it and would never, left to themselves, live anywhere else.

Part of my purpose was to visit the remaining old-timers . . . to hear their stories of old days on the Marsh and to get a sense through them, and of certain literary interpretations, of the continuous sweep of man through time and nature. There are not so many of the old ones left now. They have melted away into an eternal distance — along with the old Indians and the spirits of the animals and birds that were the models for the effigy mounds that abound in the

Marsh. But some of the old ones are living, mostly in the small Marsh towns where they inhabit snug and very clean and neat retirement homes, for they have evolved from a thrifty, neat, orderly tradition.

Eddie Lechner for example is sixty-four years old, and a very strong man. When he was young he didn't take any sass from anybody. He comes from a line of Bavarian brewmasters. His father was a highly skilled brewmaster who emigrated to the United States about 1900 to get away from army service required by the Kaiser. When his father died, Eddie took over the family tavern business in Kekoskee, a village on the edge of the Great Marsh. In prohibition times Eddie learned to be tough to survive. He made a lot of money, but he didn't keep too much of it, for personal reasons I'll relate later. He went to work for Jim Bell at the State of Wisconsin Headquarters of the Marsh, and became sort of Jim's right-hand troubleshooter, because Eddie was so expert at human relations . . . if a farmer got irate about the geese that were eating all his corn and threatened to do something violent about it, Jim sometimes sent Eddie to fix things. He was generally able to get Jim a satisfactory result. Folks at the Marsh really are fond of Eddie, and will generally do what he suggests. Jim wisely assigned Eddie to take me around the neighborhood when I was collecting the human stories for this book. We got to be good friends, and every morning when I would come up to the Marsh from Madison, Eddie would have his truck ready and we would pile in and go visiting around the countryside. He knew all the older folks, and since about all of them were German and he certainly was, Eddie was in excellent ethnic rapport with them all. Some exceptionally pleasant visits were possible only because Eddie opened the local doors.

One clear, beautiful and very cold morning when the whole Marsh sat still and brown where the cattails and the tall grasses waited in the solider places among the snow-covered Marsh ice, and a couple of hawks hunted from a sky that seemed to reflect the Marsh winter hues, Eddie and I set out for Mayville to visit with Mrs. Martha Krueger.

I never saw a dwelling as well swept and polished as Mrs. Krueger's. She lives all alone, doing her own housework at eighty-six, and occasionally she will still cane a chair, or meld the broken pieces of old china plates and saucers she has picked up at refuse heaps into an astonishing mosaic. Her entire life was spent on the Great Marsh in the hardest kind of labor. Her husband came home one afternoon in 1919 from World War I, and rural boy and girl married and went right to farming on the same place where Martha was

Frozen Waterfall Near Mayville

Marsh Guide Ed Lechner

raised as a girl . . . a farm right out on the Marsh.

Martha Krueger sits with us at her kitchen table and her eyes light with her memories. She is remembering how she, a young girl, followed her father in the wheat field that he had planted and was harvesting on the Horicon Marsh. The farm was near the east branch of the Rock River, which in those days, about 1895, was so full of fine fish — northern pike weighing up to thirty pounds, bass, pickeral, even trout — that the settlers were able at any time to catch any amount. No one was hungry on the Marsh in those days. For centuries the old Winnebago Indians and the Potawatomi had known that the Horicon Marsh was a land of plenty. And the fertile Marsh edge lands grew fine crops . . . if there was not too much water!

And the girl labored along behind her father, as he swung the heavy grain cradle, and she gathered the grain and tied it into bundles. The bundles were shocked up to await the day of threshing. Later her father got a platform reaper which didn't bind the grain either, but someone had to sit and rake the cut grain off the platform, and the girl followed and tied it again, struggling in the moist heat to keep up with the machine.

After the work was done, sometimes, not very often, a young man might come in a buggy to take Martha to a dance in Mayville, eight miles away. The Bonacks lived away back in, and the road wasn't good. Also a person had to know well how to drive the road at night, for it wasn't straight, and the nature of the land was very treacherous. There were holes that might come in the road as the wet seeped up to the surface and a wheel suddenly go down the depth of itself, to be mired; or a horse might founder into a pothole and mire down and have to be helped out. Young people in those Marsh days didn't expect to arrive at a party immaculate. More likely they were covered with mud and dirt, and often Martha Bonack carried with her a change of clothing.

To get to the Bonack place you had to go up a high slope and down another. One time a neighbor died and the small funeral procession was stalled for many hours while the hearse that carried the casket was struggled with, for it had sunk floor deep in the middle of the road. But the Marsh settlers were used to that sort of thing, and they made up for the struggle with the joy that they took in life, and in the songs and dances and the neighborly ways of helping each other, and in the mysterious presence of the great duck flocks that were everywhere and the plenty of the good earth and the fish-teeming waters.

The Bonacks lived well. Hog butchering, when the meat was prepared: salted, smoked in a special smoke house that Martha's father had built and the hickory wood he got from the Marsh timber, and cut and seasoned and used for smoking the meat. And in the kitchen her mother fried down a supply of the pork, and put it away in large crocks, or made delicious country sausages, large cakes, fried and put down in crocks in deep lard; and on the winter mornings, early when her father would go to the kitchen and start the fire in the cookstove, one by one the family would assemble in the kitchen, hugging up to the stove, and the mother would come from the pantry with a stoneware vessel of sausage and dig the hard, frozen

Harvest Near The Marsh

cakes out of the lard, and into the smoking skillet she would plunk the large cakes. The smell of them as they browned, while she made the pancake batter — buckwheat, with the starter she had kept all the winter, and the way the cakes rose, and the bubble of the brown sausage fat in the skillet and the brown sides of the light cakes, with sorghum poured over them ... it was a good way to live even though they had no real coffee. They made the coffee from barley which they roasted and ground up.

They worked very hard, though the wild game was so plentiful they never had to worry about something to eat. Wild duck got to be so common it was considered a second grade food. Mother Bonack prepared wild duck by frying it; sometimes it was roasted but much more often was just cleaned and put into the skillet. And they played hard. Father Bonack squeezed the concertina very gaily and sometimes the neighbors would come to the Bonack house for a party. The party would usually last all night, and the dancing was vigorous and very rough. Everybody could dance. The schottish, the waltz, the two-step, the polka — everyone knew them and they danced and laughed, and usually the whole neighborhood was there. There was always something to drink — it depended on who had it, and how much — but they did have homemade beer and wine and moonshine, and there wasn't any Puritan self-consciousness about using these good things, because almost all the neighborhood was German and they were folks who had been used to having beer and liquor all their lives. Nobody got drunk, but everybody had a delightful time.

One time a neighbor boy, Albert Clark, went hunting in the Marsh and he stayed out too late. He didn't come home, and finally his family knew that Albert

was lost out there. Father Bonack had been working in the fields all day and he was very tired, but when the Clarks came and told him about Albert he immediately got ready to go. Father Bonack knew the Marsh better than anyone living along it, and he led the men in to hunt. It was a very dangerous expedition, because dry-year Marsh fires had burnt deep holes into the peat and these filled with water, and were extremely dangerous. If a man got into one of these he might easily be lost, for the vegetation that had collected in the potholes was impossible to support any weight. It sank, like the surface of a bog, and whatever was on the surface went down.

Many a woman, that night, wished that her man did not have to go. But she knew that he must go. For everybody helped one another.

The boy, meanwhile, had made his way north, mistakenly thinking that he was coming out on the south edge; he came out instead away north and east at Strooks Point, and there someone found him, and sent word back that he was safe. The men in the Marsh didn't know this, however, and all night they searched. No one else was lost.

If a person is lost on the Marsh the way things are nowadays he can always find his way out by simply following one of the ditches. They will always lead him out. In those days, though, the vegetation grew very tall . . . sometimes taller than a man's head, and if one were in the Marsh and were lost, he might not be able to recognize the marks that were even very familiar to him. There was something mysterious about the Marsh, and it took hold of a person's imagination someway, and twisted his thoughts and made a kind of confusion, the old-timers said, that could dismay a mind and lead a person to have fanciful visions.

One time an old settler had become lost and confused in the Marsh and wandered around for a long while and eventually came upon his own farm, and didn't recognize his own buildings. He went to the door and asked his own wife how to get home! She didn't swing the skillet she was planning to bash him with, but saw instead that he was bewitched by the Marsh and led him into the house to introduce his family all over again.

There were two hermits who once lived in the Marsh and they lived over near the Rock River under the hill. Their names were Julius and Herman Stroede. They lived there alone in an old shanty for years and years. Herman worked quite a lot helping thresh and doing general chores, but Julius was in bad shape and didn't do much work at all. Herman always wore a lot of badges on his coat: Civil War badges and a lot of other kinds. He always walked to Beaver Dam, about fifteen miles, and once he walked clear to

the St. Louis World's Fair in 1906. To St. Louis and back! He wouldn't take a ride with anybody. He was that independent.

Martha Krueger said: I know that when Herman would walk to Mayville, he would stop off in the garbage grounds, and there he would pick up all kinds of stuff. There was a Lutheran minister in town name of Gerhardt Schrectner. Herman would agree reluctantly to take a ride with this preacher once in a while. Sadly the end of brother Julius was that once, during a cold spell, he was climbing through a barb wire fence, and he got hung up there. He couldn't get the barbs out of his clothes. And he froze, hanging there in the fence. It was awfully sad, because Julius was a good man, and he believed that man should only act natural and do just the things that God had provided for him to do. And Herman also, unbelievably, froze to death when he ran out of wood and was not able to get out to cut more.

When I was a girl I went to school, but it was always walking. I went to school, walking five miles, and in bad weather we couldn't go. But the little school was a good one. And we worked so hard all

the time, too. My father couldn't make a complete living on the unpredictable Marsh. He had to go working. The Marsh was too wet some of the time. If it was dry, then you could get the hay, but it was never knowing what would happen. But the land was good grey ground out there. We raised all kinds of vegetables. We had always nice watermelons, and good corn and oats. And we built us a stone silo so we could have silage for our cows. It was nice living out on the Marsh but there was a lot of thinking sometimes: how to make a living.

My father was a great hunter and we always had ducks to eat. Once in a while we would have a goose, too, though there weren't then so many geese as now. Mother would pick the ducks dry, and then we would nearly always fry them in a skillet. Fried duck was very, very good, because the ducks then were always fat. Quite different from the wild ducks today.

The geese always migrated through the Marsh in the spring only. Never in the fall. The geese never stopped at Horicon really, until after the refuge was established. My father when he went duck hunting, had a special suit made of hay. He looked like a straw man! He would put this long hay coat on and it reached clear down to the ground.

Out in the Marsh, especially on the islands, there was great patches of raspberries. And there were beds of wild strawberries. The wild berries were sweet, and when the children gathered them and took them to the house Mother Bonack sometimes made a shortcake, and the small, sweet berries on the hot biscuit cake with fresh cream poured on and running down and filling up the plate around the red berries — that was one of the maximum thrills of childhood. My sister and I, when we were quite small yet, had to carry the butter into Mayville for market. We had eight cows and each of us girls had a basket with some pounds of butter in them. We were little and the day wasn't very good. It was raining. We got to town, delivered our butter, and started home. But the Rock River bridge had been taken out. They were afraid the high water would wash it out. They had simply tied teams to it and pulled the bridge out! That was near our home and our father, who was afraid that in the darkness we would miss the bridge and plunge into the swift-running Rock River, was waiting there on the opposite bank for us to come. When he heard our little voices there on the far bank he shouted that the bridge was out and that we would have to walk two miles upstream to where there was another bridge. We were very small, and very tired; but we set out, and finally, hours later, we did get home. I have never felt anything so good as the way the clean sheets felt that night.

Mother Mink

When the glacier that drove so deep into the earth had carved and worried and mutilated the form of the land to create and cause Lake Michigan, there were other, smaller, inexorable mile-high barriers of ice tearing the earth to the west. Some of the scooped-out places became lakes, and through centuries of erosion and change the lakes filled with peat and became extensive wetlands. When the white men arrived about 1640 they named the wetlands to the south and west of Lake Winnebago, the Winnebago Marsh.

Before the white men arrived the Great Marsh breathed under the leavening seasons, tossing wild tall grasses and rippled waters, floating water anemone so brilliant, and the water lily on the blue, clear water. On hummocks the wild roses knew the gentle winds that also moved the snapdragons and stirred the purple vetch; and deep in the bogs the winds turned forward and backwards the yellow thrusts of marsh marigolds.

The Great Marsh was more than a hundred square miles, in shape roughly oval, surrounded by a forest of oak, maple, ash, elm, and hickory, and to the west the prairie spread widely, with oak openings where the fires had burned away the tenderer less hardy growths and where the prairie grasses waved higher, at times, than a horse's back.

The red men, trailing through the grass, were often hidden in its tallness, and their traces came down from the north and the east and up from the south;

Wild Grasses Recall Ancient Days

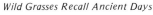

and they found their campsites where there was water and game . . . where a clear spring flowed up from the inner earth. West, from the ridge that separated the Great Marsh from the boglands of the Fox River, there were the reeds swaying in the summer winds, or crackling dry in the harder bitter winds of winter. Within the Marsh were the channels of clear water where the wild rice beat in the winds, heavy-headed, and the Indian women and children came in canoes to bend the rice stalks over and to pound the full heads into the canoe bottoms.

And as the red women beat out the rice heads, the birds among the willow and poplar, or clinging to the taller stems of the cattails, or from among the lower grasses, called with a hundred voices. The warblers of many varieties, the yellow headed blackbirds, the redwings, calling as they clung and swung; the orioles and meadowlark, and the grosbeak and the many sparrows, the thrush, and flycatcher and flicker. The roughwinged swallow and the barn swallow, and the martins and sapsuckers and robins, the vireo and kinglet and kingfisher . . . And the ducks which made the Marsh their kingdom as they followed their flyway in fall and back in spring. In the bulrushes the redhead and canvasback and the ruddy. The gadwall, the teal, the shoveler and pintail and the lesser scaup nesting in the whitetop grass or in the thistle meadow. Mallards in the reeds. In the shallows the great blue heron and bittern and lovely white egrets and swans: booming, waiting, or floating.

In the evening the sky above the Marsh clouded with endless flights of passenger pigeons, and within the hum and the almost metalic flutter and rip of the wings was the deep throat boom of great frogs at the Marsh edge.

Throughout the whole Marsh resided the muskrats. Thousands and thousands they built their low houses on nearly every wet part, and the rat sought beneath the waters for the stems of the water lily, or other juicy stalks. He stored his food in a nearby storage house connected to the main house, and he was safe throughout the winter, unless a sudden deep freeze-down blocked him from readmittance to his own dwelling and he was left to run and die on the frozen surface of the Marsh.

His enemies in summer were the large pike which waited for the muskrat shadow above and struck savagely into the shining fur with sharp, terrible teeth; or the snapping turtle taking the rat swimming and dragging him to the Marsh bottom. The mink was wild in savage killing of the rat and followed him into burrows, going wherever the rat went, and killing all the members of the rat household. The muskrats were prolific breeders. Two or three families to each mat-

Cattails Also Remember

ing each year kept the rat population on the Marsh high, until the cycle turned and thousands died and the breeding cycle diminished and for a while the rat numbers were low. Then, with the cycle turned again they came back, more than before.

On the Marsh the old Indians caught the rat for his bright skin; and when the white men came, they trapped him in greater and greater numbers, even though the price of the skins was very little . . . less than a nickel at times.

Through the Great Marsh drifted the river. The river was called the Rock by early white men, because it cut through rocky formations. The Winnebago people say that when their people and the Prairie Indians (Potawatomi) camped on the banks of the Rock River, there lived in that stream a huge and terrible water monster. This water demon the old people described as a long-tailed animal with horns on its head, great jaws and claws, and a body like a big snake. It ranged over the whole length of the stream from its mouth to the foot of Lake Koshkonong. It preyed upon both animals and men, seeming to prefer one no more than the other. Hapless deer that went to the banks of the river to drink or walked out into the water were seized and swallowed by the monster, horns and all. At the fording places of the river this demon especially hunted for victims. Indians crossing at these places were dragged down out of sight beneath the water and were never seen again. Canoes in the river were sometimes overturned by its limbs or a slap of its tail and their occupants submerged and lost. Only a few people ever saw this water monster, but its presence in the river was known by the churning and boiling of the water.

In the springtime its movement in the river broke up the ice and heaped it against the river banks. Its dens were in the deep places. There it slept and devoured its victims. Some Indians believed that there were several of these water monsters in the Rock River. Offerings of tobacco and various articles were cast into the river to appease their wrath when they were angry. These gifts preserved the lives of many people.

When the Indians ceased to camp in numbers along the Eneenneshunnuck (river of big stones) after the white men came, these water demons also apparently left the river. Bob Personius, director of the Federal Marsh, however, claims that the "Marsh Monsters" are still around, and when he speaks to groups of children from the neighboring schools he sometimes warns them to be on the lookout for strange-looking "Ogopogos" in the river and popping up of an evil head or tail in one of the ditches. I guess the kids recognize a frustrated writer in Bob and take the usual delight

Soil Study Begins

River Cleanup 1971 Headed By Mayville Jaycees

that any normal imaginative young person takes in an old folk story well told, or retold. Bob is in good company. The old story is also told about the Fox River, and how a great serpent writhed and crawled his way from North to South until he had wallowed out the channel of that important Wisconsin stream. In any case the Rock River swung easily south in two main branches and within the Marsh the east and west arteries of the river joined, and not far to the south of where they came together at a place the white men later called "The Y" the river crossed a stony ledge and made a fast rapids. It was at this rapids that the early white settlers such as Mart Rich were later to

make a dam, and very early to begin to cause dynamic changes within the life cycles of the Marsh.

Nowadays the kids pile in to help keep the Rock River free of human debris. It will probably be a while yet before the name "Horicon" comes to mean what it originally meant in "clear, pure water," but the young people are working hard to achieve it. And there have been times when the pollution problem seemed too great for even the energy and idealism of youth to overcome, when the masses of detergent foam gathered at the dam and dead fish were one result.

32

Ed Lechner, as we roam the Marsh edges, seeking tales, tells me that he has three children. One son works for an engineering company in Mayville; his daughter lives on a farm near the Marsh-edge village of LeRoy, doing well. And his third child, a son, was birth injured. That is why Ed works for the Wisconsin Department of Natural Resources. He had always been in the tavern business; his father, as he told me on our first visit, was a brewmaster from Bavaria, and was in great demand, for then there were small breweries everywhere. Ed and his wife Esther had their handicapped boy at Mayo's four different times. They took him to the spa in Waukesha for a year, and then Ed heard about a doctor at Denver. He had read quite a bit about him. But Mayo's advised against taking the boy to any more doctors. The only thing that was going to help the boy was to spend more time with him. So when they got back from Mayo's, Ed said to his wife: "Well, that's it. We'll do what he said."

And it wasn't the money either. Ed didn't regret a cent of that. He'd have spent everything he had if it had done any good. So Ed sold his tavern, and bought a place on the pond at Kekoskee, and one day Jim Bell over at the Horicon State Marsh Headquarters called him up and said, "How about coming to work for us, Ed?" So Ed asked Jim how much the State would pay for wages, and when Jim told him, Ed said that he could make about as much in a day at the tavern as the State was paying for a month's work. But Jim said, "Well, come over and try it, it's secure." So Ed went over to the Marsh and went to work. He stayed a week, then he stayed all fall, and then he was there for good. It was the work that Ed really liked, and he would be out on the Marsh a good deal of the time. And just being with his son and having a job that let him be at home at nights . . . well, it did the boy wonders. He was able to get around, and got real interested in what was going on, and just developed unbelievably. Ed thinks a whole lot of Jim Bell and the folks over at the State Marsh Headquarters.

We were driving along silent Marshland back roads in the State of Wisconsin truck and suddenly Ed began to tell about some of his early experiences. He said:

When I was a young man in the 1920s I worked for the power company in Mayville, Wisconsin. I worked on the line crew. And these were the years when the State was thinking about buying up the Marsh. The Marsh, in that period, was closed to hunting; but we ignored this. We just went out there and hunted. There was only one warden, and what chance did he have to catch us? And our power company

foreman, he was quite an outlaw, and the manager of the power and light, he was quite an outlaw too. Well, those were the days of change. The Model T was giving away to the Model A, and the old Kissel car that used to be made over at Hartford . . . it had quit business in the '30s. . . . you know, just big changes. Radio and all that had come. Well, the three of us, me and the boss and the manager, we went over one day to Mister Bill Schreiber's farm. And where the Federal dike is now, that time it was called Wheeler's Ditch or Schreiber's Ditch. And that's another story, how they built that ditch, with that great dredger that come in 1910 scooping the bottom out of the river and ruining it that way; and how Bill Schreiber, later on, a salty old German, felt when they forced the sale of his farm and scooped away his hill to build the Federal dike to hold back more water . . . but that is another story. Anyhow, we went out there to the Marsh in the early morning to do a little hunting. About four o'clock we was out there and ready.

Quarry Reflection

Huck Finns On The Rock

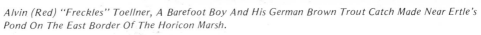

Alvin (Red) "Freckles" Toellner, A Barefoot Boy And His German Brown Trout Catch Made Near Ertle's Pond On The East Border Of The Horicon Marsh.

Well, while we was there our truck got stuck. It was a company truck we had. We weren't supposed to have that truck either. So we flipped to see which one of us would walk for a team of horses to come and pull our truck out of the Marsh. Well, I was the one who had to walk back, and that was the way it was supposed to be I guess, me being the youngest, and they being the boss and the manager. So I walks back to Bill Schreiber's and went into the barn, and there he was milking his cows. And I told Mister Schreiber what happened; and he says to me, with his head up against the cow's flank and he never stopped milkin' . . . He says, no, I will not take my team of horses up into there. He says, it is too risky. They might bog down or step in a muskrat hole. And I says, well, Mister Schreiber, when you get done milkin' are you going into town? And he says, ya, I am gonna haul the milk into town. And I says, good, so I helped Mister Schreiber with his chores, and when he gets done he pours the milk into the milk cans . . . they done that some different those days, too, and we put the cans into the back of his old Essex car. They were a somewhat unreliable rig, you remember the old Essex?

Well, we were riding into town, and I notice some little round balls on the dash of this Essex. And I am thinking, now, what the hell are them? And he is talkin' to me, and all at once he reaches over and takes one of them balls and puts it into his mouth. He starts chewing. And pretty soon, as I am watching him, and not knowing what he is doing, he lowers down the car window and spits.

And then I got on what he was doing. All them old German settlers was very conservative. What he was doing, when he had chewed a chaw of tobacco for a while, until he had got out all the first juice, he just took the chew out of his mouth and balled it up and put it on the dash. And when it had dried out good and he was going to town the next time, he just chewed it again. That's how them old boys got ahead.

Well, this old man he built his own silo out of rock he picked up around his farm. Just built a whole silo out of it, and it worked just as well as a store silo. But he saved a lot of money that way.

I try to tell these younger kids that . . . we got guys up at the Marsh headquarters, working for Jim Bell, and they got college degrees, and I don't. I only got an eighth-grade education because my dad died when I was just goin' on fourteen and I had to stop and support my mother.

Well, what I am trying to say, I appreciate how the old-timers done things. I suppose I know more about the Marsh and about the people here than anybody in our whole staff. And when there is a tour of folks who come and want to see the Marsh, well Ed Lech-

ner has got to take them.

I know all the Indian mounds on the east side of the Marsh pretty well, and I know where the old Indian trails went; and when the snow goes this spring I can take you and show you where there is an old tree with a couple of branches that the Indians bent to mark the trail, you know.

Well, to go back to how we got out of the Marsh that morning: we had quite a bit of game and we wasn't too anxious to meet the warden, so I went

Liftoff

Fall Cattail Burst

back to the office of the power company and we got another truck and some chain, and some of the crew come with me and we got the truck out of the mudhole. The next day we had a big duck feed. That's the way we done it then. Of course we couldn't get away with that today, and we wouldn't want to. But it's a different day entirely.

In a wet year these Marsh farmers had a hard time. If it turned wet before the harvest they just weren't going to get in their crops. There might be a year or two when he had a good year, then he'd have a year or two when it would be bad. But this was good in a way. Because it linked them all together. They were like one family hoping, fighting against the elements, helping one another. So they worked together and they played together, and often as not the daughters and the sons married the daughters and the sons of the neighbors; and little by little it got to be a real big Marshland family, with everybody sort of related.

They worked hard. They were rugged. They could handle their beer and liquor. There was never any Prohibition as such out in the Marsh. It didn't affect them because the next day they worked off whatever they had drunk the night before.

There were two little breweries in Mayville in those days, the Ziegler Brewing Company was one of them. And I've got an old quarter beer barrel that I found under a corncrib that was from this old Mayville brewery. And there was the Steger Brewing Company at Mayville. They both quit when Prohibition came in. They both made very good beer. In fact I think the beer they made years ago was a lot better than the beer the big companies make today because they got these laboratories today where they can cut down on hops and barley, and I heard that they even use rice nowadays. In the old days each brewery had a brewmaster from the old country, usually from Bavaria, because that was where all the great brewmasters come from. They had brewmaster schools in Bavaria, and it was in one of them that my own dad was trained.

There was a brewery in the Marsh village of LeRoy and there were two breweries in Horicon. Imagine that! That little town of LeRoy on the baldheaded landscape had its own brewery! I understand from the old-timers that the breweries around this part of the country, this country that sort of borders the Horicon Marsh, that the breweries would brew beer in the wintertime. This was because in the summertime the beer would get too hot and they might get a wild fermentation. Well, in the wintertime they would brew the beer and keg it and put it in underground cellars. I can show you where the old cellars were in Mayville and Horicon.

And when summer came the taverns would tell the

breweries how much beer they wanted and the big wagons would come with the driver sittin' away up high on the wagon seat, up above his load, you know, and with the big horses they drove, with ribbons hanging from the harness and jingle buckles going, and decorated manes, and the wagons would pull up to the taverns with a load of beer barrels while most of the town of LeRoy or Knowles would be out there watching. It was a big event when the beer wagons come.

And they got this beer out of the cellars. So they had to brew enough in the wintertime for the whole summer. They didn't have any modern refrigeration and it was kind of tough for 'em . . . 'course, they probably did have an ice cellar, where they had cakes of ice from the lakes cut during the winter, but the low cool of the cellars was just right!

Gard: That was probably pretty darn good beer.

Lechner: Yes, it was.

Gard: Your dad was a brewmaster from the old country. Do you have any of his recipes around?

Lechner: Nope. Those old guys memorized their recipes for making beer. They wouldn't tell 'em much. Dad told me, but I was about the only one. I know 'em of course.

Gard: You know much about his life in Germany?

Lechner: Well, he served three years in the army under Kaiser Wilhelm, then he got out!

Reflections In The Rock

Ice Cover

Reflections In Marshland Waters

Ed said that he himself was born in Buffalo, New York, and his brother was born in the State of Washington. I guess the way Ed's father got connected up in Horicon was through the hops salesmen who used to travel all around the country by train. They were always on the lookout for good brewmasters who might be wanting a change. If they could locate and get acquainted with all the brewmasters in the thousands of small breweries that were scattered all over the country they could increase their hops sales many times. The cooper salesmen too were always on the lookout for business. And the brewmasters, most of whom had just come over from the old country, were always looking for jobs where they could make more money, too. So the hops salesman might say, hey, there's an opening I know of out in the State of Washington. Ed's father would write there to find out about that job. That's the way he got the Washington job. Ed's dad got the job promised to him by letter, and he went out to Washington from Buffalo, and got

set up and wrote to Ed's mother to have an auction and come on out. But when they had just got nicely settled out in Washington, the State of Washington went dry; then Ed's dad got a job in Canada. This was in Alberta, and World War I came along, and the French Canadians got very suspicious of Germans, or of anybody with a German name, even, and in fact the folks all over the country were making it very hard for anybody of German descent. Especially in Milwaukee. Anyway, pretty soon the boss up in Canada told Ed's dad that he was afraid he would have to let him go. Everybody was afraid Ed's dad would put poison in the beer!

The family was escorted by the Mounted Police out of the country, and they went into Milwaukee, and there Ed's dad, by chance, met his brother from the old country. The brothers decided to go out to the Marsh village of Kekoskee and buy a mill, and Ed's dad started a tavern in with the mill and called it the Fox Head. John, the brother, who was a teamster,

40

peddled Fox Head beer, made over at Waukesha, from farmer to farmer, by the case. And since the population around the whole Marsh country was pretty much German, the Fox Head business was quite good. I might add just a word here about the water from which this Waukesha beer was made. It was sparkling "Waukesha water" from deep wells that supposedly had a real medicine value. During the World's Fair (The Columbian Exposition) in Chicago in 1893, my own mother who, a Kansas farm girl, attended the Fair, drank Waukesha water from a pipe which had been run ninety miles from Waukesha, all by gravity flow. It was one of the wonders of the Fair.

Ed's dad ran the mill and the tavern, grinding feed for the farmers, and grinding flour. Then, woe to all, Prohibition set in.

Ed's dad sold out to his brother, John, and bought the pop factory in Mayville. His idea was that he would make a little beer and a little wine; he thought the pop factory would be a very fine disguise. It worked pretty well, but he soon sold the pop factory, but he kept a tavern he had in Mayville down by the railroad track. These were the days when the iron furnace company was still going, and the coke plant and the iron mines down at Neda were going good. Ed's dad taught Ed how to make wildcat beer and how to make moonshine and wine, and they had plenty of customers at the Mayville tavern which was just going on as though nothing had happened.

Ed thought that this tavern was going to be his lifetime work. His trade. He didn't realize that this was Prohibition times. He just thought, because about everybody else in and around Mayville did, that Prohibition was just something for the rest of the country but had nothing to do with the home town. But after the tavern got going good the State and Federal men moved in and the Lechners had to pull back some from a lucrative business.

But Ed's dad died when he was about sixteen years old and Ed had to take over for the family. Well, finally Franklin Roosevelt came and Prohibition left and then Ed got back to bartending again.

But in Prohibition times, at the tavern in Mayville, they had a two-way faucet. On one side they had a barrel of near beer, and on the other a barrel of wildcat beer. On the right side of the faucet was near beer. To the left was wildcat, full-bodied and strong, made from a private Bavarian recipe that Ed's dad furnished. If a stranger came in you sold him near beer. If a fellow was a regular customer, though, you sold him wildcat. Generally he raised his glass and pronounced the benediction!

Trilliums: Spring's First

Soil Study On The Marsh Continues

Dr. Edgar Bruder And Associates Excavate An Effigy Mound At The Marsh.

The local folks around the Marsh are always conscious of ghosts . . . the shadows of the ancient Indians are present, and the Marsh people know where their primitive trails ran through, worn deeply by the feet of centuries of generations of Winnebago and Potawatomi plus the other tribes who, from time to time, inhabited the Marsh. On the Koch farm, north of Horicon, a trail ran through the Koch front pasture worn at least a foot deep. Sometimes, in the night, a Marsh farmer will think he is hearing the soft sound of moccasins, and if he is of the older breed, the ones who understood so deeply what the Marsh had been, he will think of the way things were. . . .

The Rock River meanders easily down through a gentle land that today is agricultural, with great barns and the pastures fenced, and grain fields for feed to carry the large dairy herds through winters that are easily fought now with central heat in the homes and quickly plowed out roads. The west branch of the Rock River rises near Brandon and at its source, and for forty miles as it curves and meanders south in a cautious, very non-dramatic way, it is a small stream, hardly more than twenty or thirty feet wide. Sometimes in a dry summer there is only small water in the upper river; but as it comes down to the south and west, it slowly enlarges. The east branch rises out of a complex of springs near Allenton and west and within the Horicon Marsh itself, the east and west branches of the Rock come together at the Y just above the city of Horicon, and just below this junction of the streams, centuries before a dam was thought of, unless the beaver made one sometimes in the dim past, lay a glacial dike, where Horicon city now is. The Rock had to cross this dike which in the earlier times held back the waters, so that the land all above was a great lake, not quite so large as Lake Winnebago, but very extensive. It was shallow and fairly long, and contained huge numbers of fish of all kinds. The Indians considered the lake their greatest of fishing grounds, and as, little by little, the Rock River wore down the glacial dike and the waters of the great lake lowered and lowered, the area became a marsh of about 30,000 acres . . . multitudes of migratory birds came then, and the Marsh became a chief habitation and breeding ground for geese and ducks.

The erosion stopped when the Rock River reached the hard Galena-Dolomite rock strata seven miles downstream from Horicon where the Rock comes to the Hustisford Rapids. This solid rock acted as a natural dam and maintained a fairly constant level of water in the Horicon Lake. But sediment from the adjoining streams gradually filled the lake in, and

vegetation growing rapidly, dying and growing, season after season, gradually filled in and formed peat bogs six to eight feet thick.

It took centuries for the peat to fill in the lake. It was the uniform level of the water that allowed the luxuriant and undisturbed growth of the ancient flora, particularly vast beds of wild rice and cranberries. Every kind of fish and waterfowl came in unlimited flocks and schools to the Marsh, sugar maple grew everywhere along the shores, nut and fruit trees created groves for the sustenance and pleasure of the Indians, and it was an ideal and happy place to live. If there were extensive Indian wars their record is lost in time. The livin' was just too easy. The Indians lived in real plenty, and the trails they made which were coming and going from the Marsh were worn for the purpose of obtaining food, not war.

The Woodland Mound Builders, whoever they were, for their names and origins are obscured in time, had their ceremonial spots around the Marsh, and created their great mounds in the form of mammals and birds, and serpents and turtles, and they and their inheritors left their dead buried in many locations.

Great Care With The Ancient Ones.

The old Indians left other unique signs of their habitation of the Marsh and the area. In the 1940s three adventuring boys from the Ledge School, a country school near the east side of the Marsh, left the school grounds during a noon period and climbing up along the great wall called the Ledge, or sometimes, locally, the Mayville or Horicon Ledge, they scrambled about, peering into crevices and into the overhanging rock shelters, as boys will, conjuring foxes and wolves and bears, and not giving much thought, probably, to ancient Indian cultures. In one overhang, or rock shelter into which the boys cautiously peered, they saw something on the walls not like the mere rough stone of the other shelters. What Ronald Guptill, Arlen Schellfeffer, and Sherwin Fischer were looking at were actually some very old Indian rock paintings. The boys were terribly excited and told several of their friends and several adults about what they had found, but nothing much was done about them except to expose the fact that local folks had known for a long while that there were curious pictures on the walls of some of those caves on the Ledge. Nobody seemed to have known where they really were, however, until the three school boys actually located them.

Later Dr. Edgar Bruder, a Milwaukee anthropologist-dentist came to Mayville and circulated in the area searching for Indian artifacts and mounds, and mapping many of the mounds; he, too, had heard of the existence of some cave paintings and local people told him about the three boys who recently "discovered" the pictures in the rock shelter. He later told his experiences in *The Wisconsin Archeologist*, an admirable and old Wisconsin publication. Bruder got the boys to guide him . . . they responded with tremendous enthusiasm, and they all climbed up above the Ledge School to the highest part of the cliffs of the ledge, which rise about two hundred feet above the surrounding country.

This Ledge is of pre-glacial origin; it is limestone formation and owes its existence, as it now is, to erosion. It is a part of the Niagara Escarpment and is locally considered a strange remnant, and with a little imagination can be endowed with all sorts of mysterious symbolizations. It is certainly not unusual to believe that the old Indians must have felt that painting pictures on and along the Ledge were fitting, and that the Great Spirit probably put the Ledge there for their religious uses and perhaps for their artistic endeavors.

When they arrived at the top of the cliff the boys paused to get their bearings, and then they crawled down into a winding cave about twenty feet long. The cave had a six-inch opening at the top. And near the end of the cave on the south wall, they saw the

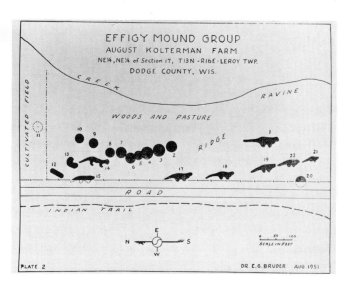

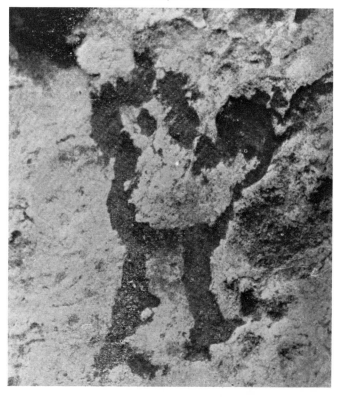

What The Boys Discovered

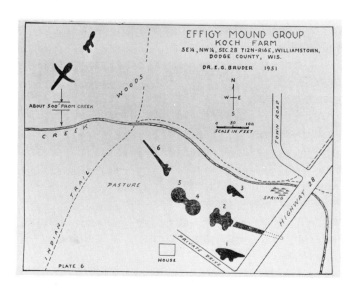

44

paintings.

The cave wall was a flat surface of dolomite limestone and the paintings were about five feet above the floor of the cave.

As Dr. Bruder gazed at the paintings he could see that they were painted in a black color with a very faint tinge of maroon at the edges, and it seemed to him that there was a sort of metallic overcast to the whole picture. The picture was of a teepee and near it the figure of a man with a bow and arrow set on the string as though he were about to loosen it. This first group showed the teepee with the poles projecting beyond the top and having a curved base. Dr. Bruder thought that the shape of the teepee might indicate that it had been left there by some branch of the Sioux Indians, since the shape of the teepee was of the kind used by the Plains Indians and not the Woodland Indians such as the Winnebago and Potawatomi which had lately inhabited the Marsh. The Indian with the bow had four feathers on his headdress and seemed to be aiming at an animal which, Bruder thought, resembled a moose. The paintings were quite small. The teepee was only three and a half inches high and three inches wide at the base. The hunter was four inches high and his bow was three inches and a quarter long. The moose, or whatever it was originally supposed to be, was only two inches overall.

While he was sketching the paintings, Dr. Bruder suggested that his young guides might enjoy exploring in an adjoining cave. The boys had been gone about ten minutes when he heard excited whoops. The scientist made his way as rapidly as he could through a narrow crevice which opened into the adjoining shelter. When he had worked himself through, he found the boys and found that they had discovered two more paintings; Dr. Bruder thought these resembled two turkeys, with tail feathers held high.

I talked with Arlen Schellfeffer, one of the boys who discovered the paintings. Arlen is now a successful farmer, still living on the old home place. He said that on the day he and the other two boys climbed up the Ledge, the furthest thing from their minds was finding any Indian paintings, but when they discovered them they all were really shocked with awe, and they all knew that they had found something really interesting. They looked hard for other artifacts, arrowheads or axes, and found nothing else that the old Indians had left.

Later, in company with some other anthropologists headed by R. E. Ritzenthaler who was then assistant curator of anthropology for the Milwaukee Museum, Dr. Bruder was able to report that Dr. Ritzenthaler had agreed that the paintings were indeed authentic; that they had been done perhaps as much as two hundred years before, and that the rock on which the paintings were placed was less soluble to water than limestone found in other sections of the country, thus the paintings had been able to survive.

The caves are not true caves. They are shelters formed by the Green Bay Lobe of the last Wisconsin ice sheet which moved, apparently, parallel to the Niagara Escarpment. The exposed rocks show polished surfaces made by the advance of the thick ice. The glaciers pushed large blocks of solid limestone over crevices and formed the so-called caves.

The whole area which is now part of the Dodge County Park, is very high, and on a fair day one may overlook the whole southern end of the Horicon Marsh, and see an area of many square miles. The cliffs overlook, too, the old Indian trails which ran to the north, along the eastern border of the Marsh.

As Dr. Bruder stood at the top of these cliffs he could see, about six miles to the north, other high ledges, where the Niagara Escarpment again came to the surface. He went later to these ledges and discovered several deep, fire-blackened holes, burnt into the rock. He thought that these holes undoubtedly were the result of many signal fires.

In the early days the Marsh was practically impenetrable. An enemy coming from the west would have had to encircle the farthest northern or southern extremities. The two signal points, the high ledges on the north and south, and the high hill standing at the edges of the present city of Mayville, could have been used by the Indians to alert an area at least sixteen miles distant.

Dr. Bruder stayed in the Horicon Marsh area off and on for a long while, and did an extensive study of the more than five hundred Indian mounds remain-

ing, even after the plow and construction work had destroyed many.

Bruder concluded that the entire east border of the Horicon Marsh area of about 50,000 acres contained many groups of prehistoric earthworks, or mounds, which generally flanked the old heavily used trails. He had much help in locating old trails and the more obscure mounds from the owners of farms who were the descendents of the original settlers, and who knew just where the mounds were located. These farmers also knew where the old Indian campsites were, and the locations of burying grounds. Bruder picked up many artifacts: axes, arrowheads, scrapers, grinding rocks for grain; and he found that many of the families in the neighborhood had collected these objects, and that several area families had notable collections, among them the Pagel family, the Wilton Erdmans, the Strooks, and others.

On a May day, rainy in 1972 I revisited the Ledge and the caves with Sherwin Fischer and Sherwin Fischer, Jr. We had trouble, but we finally located the cave and the paintings. They were very wet with the moisture on the stone; but I could make out very well the original shape of the drawings, as Bruder had described them. And then I heard that there is a great cave, plugged with tons of stone at its entrance, that has great Indian paintings on its walls.

The Trail Tree Is On This Farm

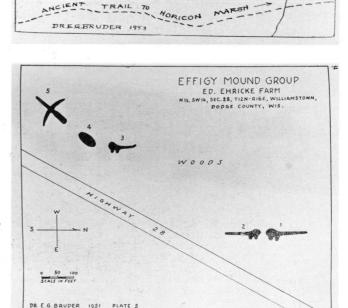

Robert Gard Searches For Ancient Indian Paintings

Among the many mysteries of the Horicon Marsh is the mystery of who did these things. Who left these imprints? I never tire of searching, with a guide who knows the whole area well, the old trails and the markers. Some time ago the ambitious and well-edited (by Ed Marolla), *Horicon Reporter,* published an account in which it was said that there were fourteen Winnebago villages around the Marsh and on the Rock River in 1820. Before this time the Winnebago had fought a war with the British that was settled by the Saint Louis Treaty of 1804. Under the terms of this treaty the Winnebago gave up none of their lands. This time of comparative happiness for them, however, was just on the edge of the great wave of white settlement that would leave most of the Winnebago lands in possession of the whites.

Apparantly John M. Kinzie, Indian agent at the Portage fifty miles away at Fort Winnebago and husband of the lively Juliette Magill who brought the first piano to Wisconsin . . . by bateaux, took an Indian census in 1828-1832. The village of Horicon was, at that time, known as Elk Village. Kinzie counted six lodges and 110 people with Old Fox as the chief. The census was taken by bringing back a red kernel of corn for each man, a yellow kernel for each woman, and a white kernel for each child.

Later Horicon was called Nech-ur-uaga, or "place of the fish eaters"; Mauk-shak-kah, or "white breast"; and Par-ach-era, or "fire village."

It must have been quite a sight that Kinzie saw when he came to the Winnebago village. The rounded lodges among the native oaks, and in the background the great Marsh with thousands of birds in the sky. The lodges were made by driving saplings into the ground about two feet apart, and then bending them together and tying them. Mats made of rushes covered this framework, and other mats were the beds. An artist's concept of this early village hangs today in the Horicon School.

Supposedly seven Indian trails met below the Horicon Marsh. The largest was the Dekorrah or Sauk Road, and led from Port Washington through the Winnebago village and so on to what is now Columbus and Sauk-Prairie. Others led from now Watertown through Hustisford to Fond du Lac; from Horicon to Hartford, Milwaukee; Horicon to Fox Lake; Horicon to Mud Lake.

The white man arrived in numbers in the 1830s, after the terror of the Black Hawk War had died away, and most of the Sauk band under Black Hawk had been slain, and the old Chief exhibited to President Jackson in Washington, D.C., where the President mildly advised the Indian and several others on exhibit to be good; for the white men would be numerous as the leaves of trees.

The Winnebago were persuaded to cede their lands to the government. The present city of Horicon was, then, built upon many Indian mounds, and during the excavation for basements, more than a score of Indian skeletons were discovered. In 1851, Wisconsin's great early man of science, Increase A. Lapham, mapped about ninety Indian mounds in what is now the middle of Horicon, and stated that these mounds were the most complicated in Wisconsin.

A friendly relationship grew between the early settlers and the Indians in the Marsh area. The Indians continued to have gardens along the east side of the Marsh, and spent winters camping in the Marsh and ice fishing. In season they harvested roots of water lillies and bulrushes, greens and wild fruit. They made maple sugar, and harvested the abundant wild rice, paddling in the Marsh and beating the rice heads into their canoes. The government, however, under pressure from the demands of white settlement, provided lands beyond the Mississippi in Iowa and Minnesota for the Winnebago and guaranteed them annuities in return. The first large migration of Indians to far places took place in 1845. There they encountered the Fox and the Sioux, who had been former enemies. The ensuing troubles caused the Winnebago to start drifting back.

But in 1850 another migration was organized. Hundreds of Indians gathered for the march at Horicon. Narcisse Juneau, son of Solomon Juneau, founder of Milwaukee, and who could speak seven dialects, was in charge for the government.

The men walked, but there were wagons drawn by oxen for the women and children and to move the canoes and other equipment from Horicon to the Wisconsin River. The Winnebago were escorted to Dakota County, Nebraska.

The memory of the Horicon Marsh with its great plenty of food was strong. Many Winnebagos drifted back to camp along the Rock River and in adjoining woods, and in 1880 the government had to remove them again. They went sadly, most of them never to return again. A few Indians lingered at or near Horicon and around the Marsh. The last Indians to live near the Marsh were the Wolf family who resided at Minnesota Junction, about three miles west of Horicon. The head of the family, Red Wolf, died in 1937. The talented daughter, Mendota, graduated from the Horicon high school.

There was always a lot of very silly misunderstand-

ing of motives between the Indians and the white settlers.

Mr. and Mrs. John Haniff who once lived south of the early village of Horicon one day saw some Indians come out of the woods near their house. Seeing them move toward their dwelling, the Haniffs rushed to their cabin, crawled into bed to "die decently." The Indians entered too, and seeing the Haniffs shivering under the scanty bed clothing, they built a roaring fire in the fireplace, made hot tea and covered the settlers with extra blankets. The Haniffs sheepily got out of bed and never again distrusted the Indians.

In 1840 or thereabouts, the Indians of the Marsh territory held a council on Badger Mound near Rolling Prairie west of Horicon. The purpose was to discuss how best to exterminate all the whites in Wisconsin. Solomon Juneau, from Milwaukee, being told of their plans, attended the powwow in person. The Indians had a high regard for him, and abandoned their plans for a general warpath.

The most famous of the Indian incidents, so-called, happened in 1861.

Early on the morning of Monday, August 26, 1861, a frantic messenger panted into the village of Horicon bringing the terrifying news that fourteen houses had been burned by the Indians at the tiny village of Kekoskee, about five miles north, and that some of the residents had been murdered in their beds. The messenger managed to impart the information that eight hundred warriors all wearing full war regalia were right then on their way to burn the village of Horicon and to work horrid scenes of rape and murder on the citizens. Everybody believed the news and the little street filled instantly with excited citizens. Crowds of women stood moaning their fate on one corner, while the braver men gathered nearby to discuss what plan to follow to protect the virtue of their womenfolk. The school dismissed, and the children ran wailing to their mothers.

Bursts Of Spring

It wasn't long until the news spread to the surrounding countryside and farmers began driving lathered horses into town, prepared for defense or a heroic death, whichever fate decreed. An effort was made to arm the men, but the arms were perhaps the most motley ever seen in preparation for a town's defense. They included pitchforks, flails, clubs picked up in the surrounding woods, a few muzzle-loading muskets, a scythe, or a sack of rocks seized along the river. Some families left for Milwaukee, and a telegram was sent from the railroad station to Milwaukee, requesting that state troops be sent immediately to Horicon to turn back the hoard of rampaging redmen. Word was sent to the village of Hustisford, about six miles south, and a company from there, armed to the teeth with clubs, rocks, and such, marched toward Horicon to help in the defense.

Shortly after noon several wagonloads of men who had gone toward Kekoskee to reconnoiter returned with the report that everything seemed quiet there. They had found a few Indians up that way, but these persons seemed to the investigating "troops" completely unprepared for war, and, in fact, thoroughly terrified at the appearance of so many ferocious armed whites. The "committee" returned to tell the inhabitants of Horicon that they need have no further fear, that there was certainly no Indian uprising, and that everyone ought to go home and forget it.

The folks at Horicon, however, would have none of this comforting advice. They regarded the small number of Indians seen at Kekoskee as a very suspicious circumstance, and they surmised that the main body of the warriors was lying doggo in surrounding woods just waiting for darkness to come so they could sneak in upon the town and begin their mission of desecration. They did, however, appoint a second "committee" to make another investigation.

The "committee" journeyed to the Indian camp. They found that the camp contained twenty-three men, with about three times as many women and children. They held a long powwow with the chief who said that his people would never fight anybody. That, indeed, if they were attacked, they would all gracefully lay down their arms, or if the white men preferred, that they would simply fold their arms and be unresistingly shot down. The chief said that he would never permit any kind of uprising. Why should they rise, so few Indians among so many white men, and why should they put in jeopardy the lush living they enjoyed at the Marsh? They had everything they wanted. They wished never to go anywhere again . . . only to live in peace.

The relieved "committee" then sat down to find out, if possible, what really had started the great Indian scare. It appeared, from long conversations with

the Indians, that there had been a few days before, a quarrel between a German settler named Dagen, and a drunk Indian. This German settler had threatened to shoot one of the Indian's ponies, and one of the ponies was actually shot, by Dagen, the Indians believed. On Sunday, August 25, one of the Indians, pretty far gone on redeye, accused Dagen of the pony shooting and chased Dagen around and round a stump, but did not ever draw his knife from his girdle. Dagen, however, sure that the Indians were going to take revenge on him, rushed to the neighbors crying, "Pytam, I vas kilt." He asked his neighbors to watch his house and haystacks for fear the Indians would burn them.

It was the neighbors of Dagen who spread the rumor, the Indians said, and the rumor grew as it traveled.

The "committee" piled censure on those whites who were in the habit of selling liquor to the Indians, and especially to those white men who visited Indian camps and insulted their females. The "committee" concluded that the lives of all the white settlers in the Horicon region were perfectly safe.

The Horicon Indian scare spread rapidly to most of the neighboring towns. At Beaver Dam, ten miles west, the Mayor received word by a special messenger that 1,500 Indians were rampaging through the countryside burning and raping. The Mayor sent a messenger to ride through the countryside spreading the word, warning farmers to flee for their lives, and many set out with their families for town, leaving house, animals, even food cooking on the stove. One old farmer drank several bottles of wine he had on his cupboard shelves to save them from the Indians.

At West Bend, twenty miles to the east, the horrible news from Horicon produced a night of terror. The excitement began on the afternoon of August 25, when the first report came of an impending Indian attack on Horicon. That evening the Milwaukee paper confirmed the news of the outbreak. About ten o'clock that night a messenger came in from the Dekorrah Road, ten miles west, crying the news that a large body of Indians was descending upon West Bend. To the wild firing of guns and roll of drums, people sprang from their beds. Children, men, women, and dogs fled in all directions. Local orators climbed on barrels to shout to the men to stand by their homes and families to the bitter end. Picket guards formed and were sent out in every direction, armed with anything they could pick up, or borrow or steal. The gunsmith was up all night repairing old muskets and pistols. Several women pleaded with their husbands to make out their wills. One woman who had been bedridden for over a year, and who lived half a mile out of town was hastily dumped into a wheelbarrow and trundled into the village for safety. A man

A Little Girl Mourns Tragedy

was reported to be shot at about 2:00 A.M. Excitement was unbearable. Then it was discovered that the fellow had accidently shot himself with an ancient pistol loaded to the muzzle with rocks. Just north of West Bend at Barton, a man stood on guard all night clad only in a shirt. He kept an ax poised over his shoulder ready to strike.

And thus, absurdly conceived and frantically carried out, was the great Horicon Indian scare, remembered with some shame for generations.

BLACK HAWK WAS THERE

Wilton Erdman, long time resident of Horicon, has a particularly deep interest in Indian lore. He is the foremost collector of Indian relics in the city of Horicon and has more than 2,000 stone relics, and bones, too, collected mostly when they were building the Horicon High School and uncovered so many skeletons of the old red men. Wilton has collected and identified more than seventy-five Indian campsites around the Marsh where he has found Indian relics.

Very early in his life as a collector, Wilton got interested in who shot the arrowheads he collected, and just who it was that used the stone axes, so beautifully shaped often, and the grinding stones for the meal they made. Wilton, through research, reading, and talking with Indian experts, found that definitely there were at least four tribes around the Marsh in later days. Besides the Winnebago and Potawatomi, the Sauk and the Fox were surely there, as well as the Kickapoo. Legends told of Sauk Chief Black Hawk having a village at the Marsh. The location of that has never really been pinpointed. The old Sauk chief led bands of Indians up from the south to the Marsh to fish and hunt, and during the Black Hawk War, in 1832, when Black Hawk and his impoverished British slanted Sauks fled North after their troubles with the whites on the lower Rock River, they definitely reached the neighborhood of Hustisford, just south of

the Horicon Marsh, and there the Indians hid for a time while the United States Cavalry hunted for them. A legend says that the old chief actually hid in an oak tree on what is now Hubbard Street, Horicon, and that the soldiers camped beneath that very tree. Another completely unsupported legend places Abraham Lincoln, a soldier in Atkinson's Army, at Horicon. But this is mere folklore. Lincoln was long gone back to Illinois, on foot, since his horse had been stolen, before the final episodes of the Black Hawk fracus.

I sought hard for somebody who could relate some Indian viewpoints on the Black Hawk War, and I couldn't find any Marsh Indians. I did, however, find old Sam Buchta, ninety-two, hunter, trapper, who lived with the Winnebago Indians on the Marsh in the old days, and whose memory includes close detail of eating with the Indians in their Marsh villages, and of hunting with them. Sam lives now in Oakfield.

Oakfield is a tidy Marshland village where English people came originally to settle. Many of the houses are old. Sam's place is more than 130 years old, and he says, has been painted only twice in all that time. Sam says that paint lasted longer in the old days, and he doesn't believe in just painting a house over and over, layer after layer of paint. Spoil the antique effect, too.

Wilton Erdman, Collector, Writer

Sam Buchta, Great Yarn Spinner

Sam recalls hearing that Chief Black Hawk of the Sauks once hid in the Marsh, though this was long, long before Sam's time, even. But some of the old Indians he knew remembered the Black Hawk story, and the Black Hawk legend persists in the whole Marsh area.

As a sidelight, Sam says that once he cured hardening of the arteries in his arms and legs with skunk oil. He took "fresh" wool, that is, wool that has just been cut from the sheep, unwashed, and soaked it in skunk oil and rubbed it on his arms and legs and left it on. The hardening of the arteries, he says, went away in two weeks. The doctors told Sam it was a miracle! Later Sam sent a small bottle of skunk oil, slightly rancid, to a lady in another state who wrote him about her own inability to move around. She did the same thing and Sam said she wrote him later saying she was back at work and doing just fine. She sent Sam two dollars for his birthday.

Ruben Gold Thwaites, once director of the State Historical Society of Wisconsin, wrote in 1892 a moving account of what happened to Black Hawk and his British-oriented Sauk band after they fled the area of the Horicon Marsh in 1832. Thwaites' story is definitely a part of the Marsh legend and tragically illustrates again what happened when Indian peoples, no matter how their motives might have been misunderstood, threatened white folks' ambitions and security. The U. S. troops had gone to Fort Winnebago and on arrival at the Fort (Portage), Thwaites wrote, the troopers found a number of Winnebago Indians there, all of them full of advice for the white chiefs. There was also at the Fort a famous half-breed scout and trader named Pierre Pacquette (Wisconsin's great strong man of legend) who had been for long a trusted employee of the American Fur Company. He informed Henry and Dodge (the troop commanders) of the true location of Black Hawk's stronghold . . . it was somewhere, he said, in the neighborhood of the Horicon Marsh.

While the division was at the Fort, during the night there was a stampede of horses from some unknown cause, the animals plunging for thirty miles through the neighboring swamps where upwards of fifty were lost.

Henry and Dodge at once resolved to return to their camp at Fort Atkinson by way of the Hustisford Rapids (where Black Hawk was supposed to be camped), and there engage Black Hawk if possible. There were delays and an attempted mutiny, for some of the troopers were not anxious to fight, but on the eighteenth of July, 1832, the troopers finally reached Rock River and found the Winnebago village at which Black Hawk and his band had been quartered, but the enemy had fled. The Winnebagos insist-

Chief Black Hawk

BLACK HAWK'S TRAIL
1832

56

ed that their late visitors were now at Cranberry Lake, a half-day's journey up the Rock River, and the white commanders resolved to proceed there the following day. They arrived at that village about noon, did not find Black Hawk, and at 2:00 P.M. adjutants Merriam and Woodbridge started south to carry word of the state of affairs to General Atkinson's camp some thirty-five miles down river. Little Thunder, a Winnebago chief, accompanied them as guide. When nearly twenty miles out, and halfway between the present cities of Watertown and Jefferson, they suddenly struck a broad, fresh trail trending to the west. Little Thunder became greatly excited and shouted and gestured violently, but the adjutants were unable to understand a word of the Winnebago tongue. Finally the Indian turned and dashed back the way he had come. The adjutants were forced to follow him, as rapidly as possible through brush and swamp. Little Thunder had returned to tell his people that the trail of Black Hawk, leaving the Marsh country in his flight to the Mississippi, had been discovered, and that there was no further use in the Winnebago trying to protect the Sauks.

The Birds Have Always Been Here . . .

. . . And The Insects

The news of finding Black Hawk's trail was received with great joy by the troopers. Their sinking spirits were at once revived and pursuit on the fresh scent was undertaken on the following morning. The course lay slightly to the north or west through the present towns of Lake Mills and Cottage Grove. Deep swamps and sinkholes were met by the army nearly the whole distance. The men had frequently to dismount and wade in water and mud to their armpits, while a violent thunderstorm with phenomenal rainfall the first night out, followed by an unseasonable drop in the temperature, increased the natural difficulties of progress. But the straggling Winnebagos, who were deserting the band of Sauk fugitives in this time of want and peril, reported the Hawk but two miles in advance, and the volunteers eagerly hurried on with empty stomachs and wet clothes. By sunset of the second day, July 20, they reached the Madison lakes, going into camp for the night a quarter of a mile north of the northeast extremity of Third lake. That same night, Black Hawk was strongly ambushed, seven or eight miles beyond, near the present village of Pheasant Branch.

At daybreak of the twenty-first, the troops were

Burdocks Eternal

up, and, after fording the Catfish (Yahara) river where the Williamson Street bridge now (1892) crosses it, swept across the isthmus between Third and Fourth lakes in regular line of battle, Ewing's spies to the front. Where today is built the park-like city of Madison, the capital of Wisconsin, was then a heavy forest with frequent dense thickets of underbrush. The line of march was along Third-lake shore to about where Fauerbach's brewery now is (was) thence almost due west to Fourth lake, the shores of which were skirted through the present state university grounds, across intervening swamps and hills to the Pheasant branch, and thence due northwest to the Wisconsin River. The advance was so rapid that forty horses gave out during the day, between the Catfish (Yahara) and the Wisconsin. When his animal succumbed, the trooper would trudge on afoot, throwing away his camp-kettle and other encumbrances, thus following the example of the fugitives ahead of him, the trail being lined with Indian mats, kettles, and camp equipage discarded in the hurry of flight. Some half-dozen inoffensive Sauk stragglers — chiefly old men who had become exhausted by the famine now prevailing in the Hawk's camp — were shot at intervals and scalped by the whites, two of them within the present limits of Madison. It was three o'clock in the afternoon before the enemy's rear guard of twenty braves under Neapope was overtaken. Several skirmishes ensued. The timber was still thick, and it was impossible at first to know whether Neapope's party were the main body of the Indians or not. The knowledge of their weakness became apparent after a time, and thereafter when the savages made a feint the spies would charge and easily disperse them.

At about half past four o'clock, when within a mile and a half of the river, and some twenty-five miles northwest of the site of Madison, Neapope's band, reinforced by a score of braves under Black Hawk, made a bold stand to cover the flight of the main body of his people down the bluffs and across the stream. Every fourth man of the white column was detailed to hold the horses, while the rest of the troopers advanced on foot. The savages made a heavy charge, yelling like madmen, and endeavored to flank the whites, but Colonel Fry on the right and Colonel Jones on the left repulsed them with loss. The Sauks now dropped into the grass, which was nearly six feet high, but after a half hour of hot firing on both sides, with a few casualties evenly distributed, Dodge, Ewing, and Jones charged the enemy with the bayonet, driving them up a rising piece of ground at the top of which a second rank of savages was found. After further firing, the Indians swiftly retreated down the bluffs to join their main body now engaged in crossing the river. It had been raining softly during the

greater part of the battle, and there was difficulty experienced in keeping the muskets dry, but a sharp fire was kept up between the lines until dusk. At the base of the bluffs there was swampy ground some sixty yards in width, and then a heavy fringe of timber on a strip of firm ground along the river bank. As the Indians could reach this vantage point before being overtaken, it was deemed best to abandon the pursuit for the night.

Black Hawk was himself the conductor of this battle, on the part of the Sauks, and sat on a white pony on a neighboring knoll, directing his men with stentorian voice.

After dusk had set in, a large party of the fugitives, composed mainly of women, children, and old men, were placed on a large raft and in canoes begged from the Winnebagos, and sent down the river in the hope that the soldiers at Fort Crawford, guarding the mouth of the Wisconsin, would allow these non-combatants to cross the Mississippi in peace. But too much faith was placed in the humanity of the Americans. Lieutenant Ritner, with a small detachment of regulars, was sent out by Indian Agent Joseph M. Street to intercept these forlorn and nearly starved wretches, a messenger from the field of battle having apprised the agent of their approach. Ritner fired on them a short distance above Fort Crawford (Prairie du Chien) killing fifteen men and capturing thirty-two women and children, and four men. Nearly as many more were drowned during the onslaught, while of the rest, who escaped to the woods, all but a half score perished with hunger or were massacred by a party of three hundred Menomini allies from the Green Bay country, under Colonel Stambaugh and a small staff of white officers.

During the night after the battle at Wisconsin Heights — as it has ever since been known — there were frequent alarms from prowling Indians, and the men, fearing an attack, were under arms nearly the entire time. About an hour and a half before dawn of the twenty-second, a loud, shrill voice, speaking in an unknown tongue, was heard from the direction of the knoll occupied by Black Hawk during the battle. There was a great panic in the camp, for it was thought that the savage leader was giving orders for an attack, and Henry found it desirable to make his men a patriotic speech to bolster their courage. Just before daylight the harangue ceased. It was afterwards learned that the orator was Neapope, who had spoken in Winnebago, presuming that Paquette and the Winnebago pilots were still in the camp. But they had left for Fort Winnebago during the night succeeding the battle, and there was not one among the troops who had understood a word of the speech. It was a speech of conciliation addressed to the victors. 60

Killdeer . . . Always Protective

Wise Owl Eyes

Neapope had said that the Sauks had their squaws, children and old people with them, that they had been unwillingly forced into war, that they were literally starving, and if allowed to cross the Mississippi in peace would never more do harm. But the plea fell on unwitting ears, and thus failed the second earnest attempt of the British band to close the war. As for Neapope, finding that his mission had failed, he fled to the Winnebagos, leaving his half-dozen companions to return with the discouraging news to Black Hawk, now secretly encamped in a neighboring ravine north of the Wisconsin.

The twenty-second of July was spent by the white army on the battlefield, making preparations to march to Blue Mounds for provisions. It was discovered that the enemy had escaped during the night across or down the river, and it was thought that the troops were insufficiently provided with food for a long chase through the wholly unknown country be-

yond the Wisconsin river.

On the twenty-third, Henry marched with his corps to the fort at Blue Mounds, and late that evening was joined by Atkinson and Alexander, who, on being informed by express of the discovery of the trail and the rapid pursuit, had left the fort on the Koshkonong, officered by Captain Low, and hastened on to the Mounds to join the victors. Atkinson assumed command, distributed rations to the men, and ordered that the pursuit be resumed.

On the twenty-seventh and twenty-eighth, the Wisconsin was crossed on rafts at Helena, then a deserted log village, whose cabins had furnished material for the floats. Posey had now joined the army with his brigade, and all of the generals were together again. The advance was commenced at noon of the twenty-eighth, the four hundred and fifty regulars, now under General Brady — with Colonel Taylor still of the party — in front; while Dodge, Posey, and Alexander

Owl Eyes Looking Back

followed in the order named, Henry bringing up the rear in charge of the baggage. It appears that there was much jealousy displayed by Atkinson, at the fact that the laurels of the campaign, such as they were, had thus far been won by the volunteers; and Henry, as the chief of the victors at Wisconsin Heights, was especially unpopular at headquarters. But the brigadier and his men trudged peacefully on behind, judiciously pocketing what they felt to be an insult.

After marching four or five miles northeastward, the trail of the fugitives was discovered trending to the north of west, towards the Mississippi. The country between the Wisconsin and the great river is rugged and forbidding in character; it was then unknown to whites, and Winnebago guides were almost equally unfamiliar with it. The difficulties of progress were great, swamps and turbulent rivers being freely interspersed between the steep, thickly-wooded hills. However, the fact that they were noticeably gaining on the redskins constantly spurred the troopers to great endeavors. The pathway was strewn with the corpses of dead Sauks, who had died of wounds and starvation, and there were frequent evidences that the fleeing wretches were eating the bark of trees and the sparse flesh of their fagged-out ponies to sustain life.

On Wednesday, the first of August, Black Hawk and his now sadly depleted and almost famished band reached the Mississippi at a point two miles below the mouth of the Bad Axe, one of its smallest eastern tributaries, and about forty north of the Wisconsin. Here he undertook to cross; there were, however, but two or three canoes to be had, and the work was slow. One large raft, laden with women and children, was sent down the east side of the river towards Prairie du Chien, but on the way it capsized and nearly all of its occupants were drowned.

In the middle of the afternoon, the steamer *Warrior,* of Prairie du Chien, used to transport army supplies, appeared on the scene with John Throckmorton as captain. On board were Lieutenants Kingsbury and Holmes, with fifteen regulars and six volunteers. They had been up the river to notify the Sioux chief, Wabasha — whose village was on the sight of Winona, Minnesota — that the Sauks were headed in that direction. As the steamer neared the shore, Black Hawk appeared on the bank with a white flag, and called out to the captain, in the Winnebago tongue, to send a boat ashore, as the Sauks wished to give themselves up. A Winnebago stationed in the bow interpreted the request, but the captain affected to believe that an ambush was intended, and ordered the Hawk to come aboard in his own craft. But this the Sauk could not do, for the only canoes he had were engaged in transporting his women and children over the river, and were not now within hail. His reply to

that effect was met in a few moments by three quick rounds of canister-shot, which went plowing through the little group of Indians on the shore, with deadly effect. A fierce fire of musketry ensued on both sides, in which twenty-three Indians were killed, while the whites suffered but one wounded. The *Warrior,* now being out of wood, returned to Prairie du Chien for the night, the soldiers being highly elated at their share in the campaign.

During the night a few more savages crossed the river; but Black Hawk, foreseeing that disaster was about to befall his arms, gathered a party of ten warriors, among whom was the Prophet, and these, with about thirty-five squaws and children, headed east for a rocky hiding place at the dells of the Wisconsin, whither some Winnebagos offered to guide them. The next day, the heart of the old man smote him for having left his people to their fate, and he returned in time to witness from a neighboring bluff the conclusion of the battle of Bad Axe, that struck the death blow to the British band. With a howl of rage, he turned back into the forest and fled.

The aged warrior had left excellent instructions to his braves, in the event of the arrival of the white army by land. Twenty picked Sauks were ordered to stand rear guard on one of the high bluffs which here line the east bank of the Mississippi, and when engaged, to fall back three miles up the river, thus to deceive the whites as to the location of the main band, and gain time for the flight of the latter across the stream, which was progressing slowly with but two canoes now left for the purpose.

Atkinson's men were on the move by two o'clock in the morning of August two. When within four or five miles of the Sauk position, the decoys were encountered. The density of the timber obstructing the view, and the twenty braves being widely separated, it was supposed that Black Hawk's main force had been overtaken. The army accordingly spread itself for the attack, Alexander and Posey forming the right wing, Henry the left, and Dodge and the regulars the center. When the savage decoys retreated up the river, as directed by the chief, the white center and right wing followed quickly, leaving the left wing — with the exception of one of its regiments detailed to cover the rear — without orders. This was clearly an affront to Henry, Atkinson's design doubtless being to crowd him out of what all anticipated would be the closing engagement of the campaign, and what little glory might come of it.

But the fates did not desert the brigadier. Some of Ewing's spies, attached to his command, accidentally discovered that the main trail of the fugitive band was lower down the river than where the decoys were leading the army. Henry, with his entire force, there-

Old Marsh Ditch

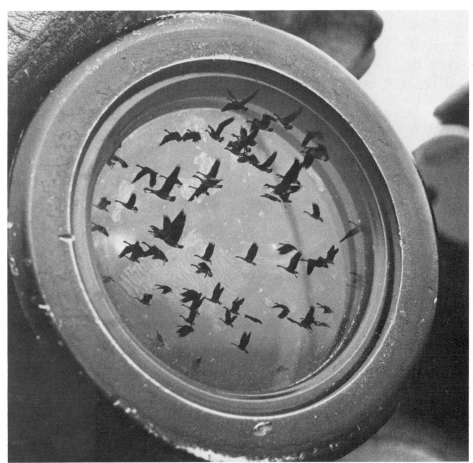

Reflections In Binoculars

Early Birds

upon descended a bluff in the immediate neighborhood, and after a gallant charge on foot through the open wooded plateau between the base of the bluff and the shore, found himself in the midst of the main body of three hundred warriors, which was about the number of the attacking party. A desperate conflict ensued, the bucks being driven from tree to tree at the point of the bayonet, while women and children plunged madly into the river, many of them to immediately drown. The air was rent with savage yells and whoops, with the loud cries of the troopers as they cheered each other on, and with the shrill notes of the bugle directing the details of the attack.

It was fully half an hour after Henry made his descent, when Atkinson, hearing the din of battle in his rear, came hastening to the scene with the center and right wing, driving in the decoys and stragglers before him, thus completing the corral. The carnage now proceeded more fiercely than ever. The red men fought with intense desperation, and, though weak from hunger, died like braves. A few escaped through a broad slough to a willow island, which the steamer *Warrior,* now re-appearing on the river, raked from end to end with canister. This was followed by a wild dash through the mud and water, by a detachment of regulars, and a few of Henry's and Dodge's volunteers, who ended the business by sweeping the island with a bayonet charge. Some of the fugitives succeeded in swimming to the west bank of the Mississippi, but many were drowned on the way, or cooly picked off by sharp-shooters, who exercised no more mercy towards squaws and children than they did towards braves — treating them all as though they were rats instead of human beings.

This "battle," or massacre, lasted three hours. It was a veritable pandemonium, filled with frightful scenes of bloodshed. The Indians lost one hundred and fifty killed outright, while as many more of both sexes and all ages and conditions were drowned — some fifty only being taken prisoners, and they mostly women and children. About three hundred of the band crossed the river successfully, before and during the struggle. The whites lost but seventeen killed and twelve wounded.

Those of the Sauks who safely regained the west bank were soon set upon by a party of one hundred Sioux, under Wabasha, sent out for that purpose by General Atkinson, and one-half of these helpless, half-starved non-combatants were cruelly slaughtered, while many others died of exhaustion and wounds before they reached those of their friends who had been wise enough to abide by Keokuk's peaceful admonitions and stay at home. Thus, out of the band of nearly one thousand persons who crossed the Mississippi at the Yellow Banks, in April, not more than one hundred and fifty, all told, lived to tell the tragic story of the Black Hawk War — a tale fraught with dishonor to the American name.

The rest can soon be told. On the seventh of August, when the army had returned to Prairie du Chien, General Winfield Scott arrived and assumed command, discharging the volunteers the following day. Cholera among his troops had detained him first at Detroit, then at Chicago, and lastly at Rock Island, nearly one-fourth of his force of one thousand regulars having died with the pestilence. Independent of this, the American loss in the war, including volunteers and settlers killed in the irregular skirmishes and in massacres, was not over two hundred and fifty. The financial cost to the nation and to the state of Illinois aggregated nearly two millions of dollars.

On the twenty-seventh of August, Chaetar and One-eyed Decorah, two Winnebago braves who were desirous of displaying their newly inspired loyalty to the Americans, delivered Black Hawk and the Prophet

Hoarfrost

into the hands of Agent Street, at Prairie du Chien. They had found the conspirators at the Wisconsin river dells, above the site of Kilbourn City (now Wisconsin Dells).

On the twenty-first of September, a treaty of peace was signed at Fort Armstrong; and Black Hawk, the Prophet, and Neapope — who had been captured later — were, with others, kept as hostages for the good behavior of the small remnant of the British band and their Winnebago allies. They were kept through the winter at Jefferson Barracks (now Saint Louis), and in April, 1833, taken to Washington. They stayed as prisoners of war in Fortress Monroe until June four when they were discharged. After visiting the principal cities of the east (and President Jackson), where Black Hawk was much lionized, and given an adequate idea of the power and resources of the whites, the party returned to Fort Armstrong, where they arrived about the first of August. Here Black Hawk's pride was completely crushed, he being formally transferred by the military authorities to the guardianship of his hated rival, Keokuk. This ceremony the fallen chief regarded as an irreparable insult, which he nursed with much bitterness the remainder of his days.

The aged warrior, with the weight of seventy-one years upon his whitened head, finally passed away on the third of October, 1838, at his home on a small reservation set apart for him and his personal followers, on the Des Moines river, in Davis county, Iowa. In July of the following year (1839), an Illinois physician stole his body from its grave. Complaint being made by Black Hawk's family, Governor Lucas of Iowa caused the skeleton to be delivered to him at Burlington, then the capital of that Territory, in the spring of 1840. The seat of government being moved to Iowa City later in the year, the box containing the remains was deposited in a law office in the latter town, where it remained until the night of January 16, 1853, when the building was destroyed by fire — End of the Black Hawk story!

Backward now from the horrible events of the Black Hawk Marsh-related War to a gentler commentary on the Indians of the Marsh.

A Sauk and Fox amalgamation originally took place at Horicon. And with these two tribes there were Menomini, the Prairie Potawatomi, and the Winnebago. The Winnebago ruled the Rock River waterway for over two hundred years.

When Saterlee Clark, the first white man to leave a report of what he saw at Horicon Marsh, went through the area in 1830, he counted enough Indians from the wigwams set up there to make up, he thought, a population of about two thousand. His account is published in the *Dodge County History.*

White Breast, Maunk-shak-kah, the Indians called it, wrote Saterlee Clark, in about 1880, looking backward, was for many, many years — I don't know how long — a noted Winnebago village. On the night of September 2, 1830, I slept in an Indian lodge on the east bank of Rock River, where Horicon now stands. There were two rows of lodges extending several rods north from a point near where the Milwaukee and Saint Paul bridge spans the river. The population of White Breast, I should judge, was close upon two thousand bucks, squaws and papooses. I was on my way, in company with White Ox, to an Indian settlement at the head of Lake Koshkonong. I was but fourteen years of age, and lived with my father at Fort Winnebago. The Indians treated me well, and I have no cause to complain of ill usage at their hands at any time during the seventeen years thereafter that I traded with them. They always possessed and exhibited the warmest friendship for me, and now, when the few scattered remnants of the once powerful tribes that inhabited Southeastern Wisconsin come to Horicon, they never go away without paying me a visit. As an illustration of their fidelity toward me, I will relate an incident that occurred a few years since. While going to Milwaukee, half a dozen Indians got on the rear platform of the car in which I was sitting with two or three ladies I met on the train. Just as we were pulling out from the station, I heard an unearthly yell, and, looking up, saw those Indians coming down the aisle on a run, throwing up their hands and uttering all manner of joyous exclamations in their own tongue. In a moment, they were upon me, pulling my clothing, shaking my hands and arms, and jabbering away with all their might. Every one in the car was frightened nearly to death. The ladies with whom I had been conversing almost fainted. It was not until they saw me shaking hands with my old friends that they recovered from their fright. There were some *pale faces,* sure enough, in that car. In the midst of the excitement, the conductor came along and ordered the Indians to "get out," but I told him it was only a little peace powwow, and that, when they got through talking, I would send them away, which I did after shaking each of them by the hand again and wishing them good luck.

How did the Wisconsin Indians pass their time? Oh,

Outdoor Classroom On The Marsh

very easily indeed, and pleasantly withal. Hunting and fishing and trading were the chief pursuits of the males. The squaws devoted their attention, during the spring and summer months, to raising corn, and the autumn and winter to dressing deer hides, making moccasins and building fires in their wigwams. During warm weather, they lived in lodges built of white cedar bark. Within these lodges were constructed, of poles and grass mats, very comfortable berths, where the weary huntsman stretched himself in sleep at night. In the winter, wigwams were substituted for their airy lodges. The wigwams were made of heavy mats prepared from the grass which grew upon the marshes and the borders of the lake. A strip of matting, two or three feet wide, would be stretched around the bottom of a series of poles placed in the ground certain distances apart, coming together at the top eight or ten feet from their base. An embankment of snow, or earth, if the former did not exist in sufficient abundance, was then thrown up about the outside of the matting; another and another strip of the same grass material being placed above the first until the circular wall became of sufficient height to protect the inmates fromthe chilling blasts of wind which howled through the forest, the top of the wigwam being left open to allow the smoke to escape from the fires, around which the Indians gathered at night to relate their deeds of war or tell their tales of love. When the drowsy god of sleep asserted himself, they would wrap themselves in their blankets, turn their feet to the fire and obey his commands. Their bed was the cold, solid earth; their sheets, simple grass mats. The couch was not downy, but it was comfortable.

Yes, they buried their dead above ground. Along the banks of the river could be seen the last resting-places of many good Indians. When one of their number died, a rude platform was constructed of poles brush, six or seven feet from the ground. The corpse, being placed in an old canoe covered with bark and hermetically sealed with tamarack gum, was then deposited upon this platform, and the last sad rites were over.

The Indians lived under the Chief, White Breast, whose Indian name was Maunk-shak-kah, same as the village. The many Indian trails leading to Horicon showed that the place was extremely well known, and that it was well adapted for trade, the Indians walking, or with horses pulling travois — two sticks fastened on each side of a pony to make a kind of sled. The Menomini and the Potawatomi are Algonkians as are the Sauk and Fox. But the Winnebago belong to the Sioux tribe. The Winnebago and the other tribes around the Marsh could not understand each other's language, but the Algonkian peoples could communicate.

Wilton Erdman says that there may also have been parts of the Illinois tribe at the Marsh, the Kickapoo were surely there . . . another Algonkian tribe. The most baffling tribe, however, and the one most shrouded in deep mystery, were the Woodland Effigy Mound builders.

One of the large groups of effigy mounds at the Marsh, called the Mieske Mound group, and which is situated between Horicon and Burnett, was partly excavated by the anthropologist W. C. McKern. There were about eighty mounds in this group. McKern spent an entire summer working at this group and about half of the mounds were excavated. He discovered very interesting information about the ancient people who built the curious mounds.

He learned that these people used no cradle boards, as did other Indian tribes, and he discovered this from the skulls of the skeletons excavated which showed no rear skull deformations. Cradle boards carried on the back caused this deformity. Animals were buried with the bodies . . . pet dogs, usually. Altars were used, apparently for ceremonial purposes. Clay or pebble cysts or chambers were often erected. Pottery, in pieces, of course, was found buried with the skeletons. Bone harpoon points, awls, were found in the mounds; curiously, McKern found only a few arrow points with the bodies . . . these were triangular in form. There was absolutely no evidence of white man's influence discovered in any of the mounds, which, McKern said, were certainly built long before the advent of the white man to Wisconsin in 1634, when Jean Nicolet beached his canoe at Red Banks (or thereabouts) with pistols blazing and clad in Oriental robes, to greet, as he thought, an Oriental people.

Wilton says that when Columbus "discovered" America in 1492 there were probably about a million Indian people. In 1970 there were upwards of a half million Indians . . . and many, if not most, of these are less than one-half Indian blood. Of the once numerous Menomini tribe, for example, there sadly remain only twelve full-blooded members.

Wilton for many years made regular expeditions to search all through the Marsh area for Indian relics. Plowed fields were the best places to look. Especially after a rain.

I asked Wilton why he spent so much time looking for Indian relics. And he answered:

I didn't much like to play golf. Hunting relics is fine exercise. And there is always that challenge. That little bit of mystery. Good exercise, too. First time I

Fish Stories Aren't All Big

collected was in 1913. I looked down at something. Right here in town, and at first I thought it was a leaf. But it was a rose quartz arrow point. I had been sent out to get some tomatoes for Mamma Gesner's Hotel that she ran on Main Street, Horicon . . . called it the Horicon House. When I got out to the tomato patch, I was with my friend Ervin Dugan; he found something, and was looking at it kind of funny, and I says, what you got? He says, I think I got an Indian arrowhead. Ervin knew what they were. Then I looked around and I found two. The first one I found looked like a leaf. That was the first great thrill I guess I ever had. Something just sort of bursted inside me. And from that very moment on, I was a collector. I started to read about the Indians, and I got further into it and then I was hooked. I couldn't find maximum pleasure doing anything but hunting relics.

I chased around all over this part of the country.

Spring is a good time to hunt, after the snows have melted, then the relics get exposed, or after a rain as I said, that will wash them out, too. You get an eye for looking. I have found many fragments of pottery. But it is very hard to piece together a whole pot. If you find the rim pieces or the bottom pieces and maybe some from the middle section, then you can pretty well tell the dimensions of the pot. I have spent hundreds of hours trying to reconstruct pots.

The Indians liked this Marsh place because their food, clothing and shelter wants were centered here. In those old days there was also plenty of pure water. When I go looking for Indian relics I always go next to the water. That's my secret, if I have one. Never wander away too far from water.

Two years ago I found a copper spear over east by the Horicon water tower. The spear point was about five inches long. I found the copper spear there, and nothing else. You would have thought, maybe, there would be some other things there too. But there weren't. That spear is very, very old. We don't find many copper specimens here. Maybe the Copper People didn't get this far south. The Fox River Valley is rich in copper. But they didn't seem to have come down in here much. To harden the copper the Indians simply pounded it. Pounding copper will make it very hard, and a copper knife pounded will have a good cutting edge.

In prehistoric times . . . the times of the Mound Builders, the Marsh area was the natural habitat of the deer. There was swamp, forest, and prairie. West of the Marsh was the prairie. The prairie commenced about a mile west of the city of Horicon, and the prairie went all the way to the edge of where the city of Beaver Dam now is. It was a wonderful prairie . . . high grass, unbelievably high, and so thick, with the knife edges of the grass sawing against legs and bodies, and sometimes so high that walking men couldn't be seen at all. And in the grass and timber spots lived the animals. There were bear, and a lot of them. And elk, and there were some buffalo, too. Maybe not so many buffalo. There was plenty of large game and small game. And the Indians had gardens. Up till about thirty years ago there were some corn hills, that could still be seen right here in the city of Horicon. They lived very well. They had everything they needed. Right here.

There wasn't much fighting among these Indian peoples. The Winnebago people submitted to the incursions of the Algonkian groups. In times before 1600 or so, all of Wisconsin, to the shores of Lake Michigan, was owned by the Sioux. But as the Algonkian peoples came in over the north passageway and they came up from the south they made a kind of pincer movement to drive the Sioux away toward the Mississippi. The Winnebago were mostly on the eastern side of Lake Winnebago and down to the Horicon Marsh area; and when the Algonkian peoples drove the Sioux away, they seem to have forgotten the Winnebago. Just left 'em!

The Winnebago didn't like the Sauk and Fox too well, but I guess the livin' was just too easy up at the Marsh. So far as I know, nothing major in the way of a war took place right around here. There was plenty for all.

There was a woman lived at Horicon in the early days, in the 1840s, named Mrs. Beers. She lived on the east side of Rock River, and she could hear the Indians and their ponies, with their bells, ringing all night.

I will give you a sketch of Horicon as it was nineteen or twenty years ago, wrote Mrs. Beers about

Fast Freeze On The Marsh

1865 for the *Dodge County History*. This place was wild, yet beautiful. It was formerly an Indian planting ground, and many of their corn hills are still visible, as they planted in the same hills each year without plowing as our farmers do . . . the Rock River flowed quietly along, and on its eastern bank near the old depot lay scattered along a number of mounds; whether thrown up by the God of Nature of the Indians, I know not, but we called them Indian mounds. They were similar to each other, usually with a large tree in the center of each. On the bank near the river, was an Indian trail worn deep into the earth; for it had been trodden by Black Hawk and his tribe, as well as other tribes for many long years.

72

Early Morning On The Marsh

There was a fine spring on the bank of the river under a large tree; it was a splendid place, and for a long time we got all the water from there that we used for drinking or cooking purposes, crossing the river in a small boat to obtain it. I well remember the first night I ever stayed in this place. It was dreary enough. The Indian ponies were grazing around the house all night, and their bells kept up a constant tinkling. The fear of the Indians troubled me somewhat, as my thoughts would go back to the narratives I had read of Indian cruelties to frontier settlers; but here they seemed harmless. They called themselves Potawatomis or Menominis, and seemed ashamed to be called Winnebagos, as the latter were considered by the whites to be much more cruel than the former. We could usually distinguish the Winnebagos by their red blankets, while the other tribes wore white or blue. Ofttimes, while about my work, I would look up to see black faces peeping in at my window, as that was their habit always before entering a house. Such sights startled me at first, but I became accustomed to it and learned to trade with them, and buy venison, fish, ducks, and berries. They usually wished to "swap" for flour or salt. They always seemed much pleased with the salutation *bazhu* from us, which means, "How do you do?" and would respond with the same word immediately. I had nothing stolen by them except a pie, which I left out to cool; but they were almost always begging, saying they had no *shoneau* (money). In the winter the squaws and papooses would come to warm themselves by our fires. I very much disliked to let them in, as the house would smell of smoke for hours afterward; but they would say, "Heap cold," and look so wretched that I could not say *puckachee* (go away). During certain seasons of the year, in the winter, there would be a large number of wigwams where the old depot now is. We visited them, and found some of the squaws dressing fish, roasting venison or tanning deer-skins. Everything seemed quite comfortable, excepting the smoke from the fires built in the center of each wigwam, which scented the surroundings with an unpleasant smell. They often buried their dead above ground, usually in an old canoe, supported by four posts and covered over with bark, but as the white people came and settled here the Indians were induced to discontinue the habit, and also to remove the old sarcophagi. Occasionally the Indians would have a powwow, which I always dreaded very much, as at such times they bought whisky, or *goodnatush* as they called it, from enterprising dealers in Grubville, now Beaver Dam. They would keep up their fires and make night hideous with their yelling, singing and dancing, but such things did not occur very often.

But the Indian story of the Marsh was only a low-key prelude to the sad tale of the white man and of his litigation and sometimes colorful attempts to develop the Marsh and to realize profit from it. The old Indians sought no profit beyond their food and what they required for lodging. Even in the midst of the plenty provided by nature in the Marsh, they did not waste. They used, they worshipped and celebrated in their own way, but they did not consciously spoil or change. Not so the white man.

The dike that nature, in the guise of the glacier, threw across the river was an open invitation to industrial development. The early settlers foresaw, as early as the 1830s when Solomon Juneau, the founder of Milwaukee, visited the site of Horicon, that the rapids at the point just below where the east and west branches of the Rock came together was a magnificent location for a power dam. If a dam might be built there where the glacial dike naturally backed up the waters, then the waters on the Marsh might be raised to almost any height. The Marsh lands, which nobody considered to have any value for agriculture, could then be flooded and an immense reservoir created which would furnish water for mills, and for industry, a resource unlike anything else in the new Wisconsin territory.

By 1839, the first dynamic manifestation of a profit element was present. David Giddings and Albert G. Ellis of Green Bay, in 1839, secured a strip of land of about 300 acres stretching for a mile along both banks of the Rock River where the rapids quickened. Then Giddings, whose partner was an enterprising early Wisconsin figure named Moses Strong, got the territorial legislature to authorize a dam at any point selected by the owners. The legislature did specify, however, that a lock should be maintained to make the River always navigable, and that there must be a chute or slide to permit the passage of fish.

Youth Views Exposed Dam During River Drawdown

It turned out that these early developers did not actually build the dam, but Moses Strong, who was agent for an ex-governor and senator from New Hampshire named Hubbard, bought the land of Ellis, and Hubbard put up the actual money to pay for it. Hubbard finally disposed of his land to a business group of three individuals, one of whom was that imaginative and colorful Horicon settler, Mart Rich. It was this group of three who finally built the dam and erected a sawmill. They followed the sawmill with a gristmill in 1848, the year that Wisconsin became a state, and soon there were other manufactories using the same water power. The dam itself wasn't a particularly good one . . . some posts, driven in endwise, some mud and stone . . . was about all, but it held the water back temporarily until in 1852, a big head of rainwater washed the old dam out, flooded quite a bit of land, and ruined some crops. Settlers banded immediately to restore the dam, and built it much stronger that time, and the settlers living in and around Horicon considered that they were luxuriating in the middle of a blooming Paradise, and that they were to become the inheritors of a new Canaan where everyone could easily become rich either from the manufacturing of products using the great new water power, or else from commerce generated by the magnificent great lake that the dam had created out of a seemingly worthless marsh. The dam had a head of water of over nine feet, and Horicon Lake, so-called, the result of back-up waters of the Rock River, covered nearly fifty square miles. The lake was about fourteen miles long and nearly six miles wide . . . one of the great lakes of the state, and by far the largest artificial one. In fact, some said that Lake Horicon was the largest artificial lake in the world.

Visitors to the lake were astonished at the numbers of fish that the lake contained. There were smallmouth and largemouth bass, pike, pickerel, bullheads, muskellunge, perch, sturgeon . . . but no carp, since it was not until about twenty years later that carp were introduced into the Rock River. Carloads of fish, pike and pickeral especially, were shipped by ice fishermen to Chicago and Milwaukee. Wild fowl, usually ducks, but not so many geese, appeared in literally uncountable numbers. Muskrats multiplied so rapidly that they appeared at times about to overrun the entire region . . . and the lake was large enough to accommodate fairly large steamers.

No story of Lake Horicon is complete without mentioning navigation. During the first years of the lake's existence, the crafts were propelled by man power. In 1857 Cyrus Gates built a small craft similar to a steamboat in structure but without the steam. It could carry twenty-five people, was driven by a lever worked by one or two men, and ran about eight miles an hour. Captain Clausen owned a sailboat which seated twenty people.

The following year, when Captain Judd operated a steamboat large enough to carry freight, passengers or to tug down fleets of saw logs from the upper waters, the residents of Horicon and vicinity were jubilant. The local weekly paper buoyantly declared: "Go 'way with your little places like Fond du Lac, Oshkosh, Green Bay, Sheboygan and Milwaukee. We have a lake and a steamer besides three railroads to help us on to glory and victory."

On the steamer's first trip to Kekoskee, people crowded the shore and shouted and swung their hats for joy. It was the first time a steamboat had ever traveled from Horicon lake up the Kekoskee river. At the Indian settlement, an army of Indians came out

Sunset On The Marsh

for a look at the queer craft. As it passed, they gave one unearthly whoop and ran for the woods.

The largest steamer on Lake Horicon was the M. Winter. It was built in the winter of 1859 and launched the following June. She was a ninety-ton boat, ninety-eight feet long and twenty-two feet wide. The hold was three and a half feet deep. She had an eighteen-inch draught and was propelled by a twenty-five horsepower engine. The cabin was forty-five by fifteen feet. She cost $4,000. Seventy-five couples dedicated it with an excursion around the lake and with a dance on it in the evening.

Michael Winters, owner of the steamer, made daily trips to Kekoskee, two miles from Mayville, and then up the lake to Leroy and Chester, a distance of about fifteen miles. Two piers about four miles apart on the east shore of the lake made it possible for the farmers in the adjacent country to bring their produce to the lake and have it carried to Horicon to be shipped on the railroad. The steamer ran regularly in connection with the railroad schedule during the navigation season.

August 17, 1859, the boiler of the steamer blew up as she was coming down from Chester with a load of stones. The boat had been stopped a few miles from the head of the lake opposite Schoemann's point to repair a rudder. In less than five minutes the boiler burst, killing two men. The boat was floated down to Horicon and repaired the following winter. Mr. Winter continued to use it until about 1867 when the boiler and engine were put in a seeder factory. In fact the old engine was the first "Power" for the Van Brunt "seeder drill," which eventually became part of the John Deere Implement Company, and exists in Horicon today. It was the M. Winter steamer that was chartered to take an investigating committee to the Indian camp near Smutt's Point at the time of the "Big Indian Scare," August 30, 1861. As related, the committee reported the "scare" a hoax.

But the lake itself was a beautiful body of water. Its shores were "graceful, sinuous, wooded and contained deep coves and headlands and there were beautiful, small wooded islands." It took nearly three years for the entire basin to fill with water and as it filled, great changes occurred at the lake bottom. The transformation from marsh to lake actually floated hundreds of acres of reedy bottom to the surface, and these slowly passed down the River and flowed over the dam. The bottom of the lake then became, or was left, sandy, hard, and gravelly.

Hunting lodges appeared along the shores of Lake Horicon. Rafts of logs, cut along the margins of the lake, floated down to sawmills at the lower end. The lake became famous for excursions, for sport, and was considered by some observers to be superior to the larger and more northerly Lake Winnebago for sheer beauty, fishing, hunting, and just pleasure outings.

The beauty and promise of Lake Horicon was not to last for very long. Ever since the dam was constructed there had been unhappiness among some area farmers who did not like the flowing of their agricultural land. In 1858 the first litigant brought suit against the owners of the dam, a concern then called The Horicon Iron and Manufacturing Company. It would have seemed likely that litigation might have started sooner, for the first suit was filed more than ten years after the dam had been constructed, and ten years after the lake had attained maximum depth. The suit, however was successful, and the Wisconsin Supreme Court in 1861, on an appeal from Dodge County, upheld the decision of the circuit court which had awarded the plaintiff damages amounting to $860 and costs.

Joseph Schafer, once director of the Wisconsin State Historical Society, states that this first decision was the critical case because it was the first legal attack upon the dam at Horicon. Many land owners around the edges of the lake were awaiting the outcome of that suit, for it would indicate whether or not it would be worthwhile for them to follow the course of the first successful lawsuit.

Suits of course multiplied, and several of these also reached the Supreme Court. At last Mart Rich gave in. He didn't have the capital to fight the lawsuits and the dam was condemned. In December, 1868, the sheriff of the county came armed with a writ to tear out the Horicon Dam. One newspaper account said that forty men, with picks and shovels, were on hand to aid the officer, but a delay was once again secured. However, in the spring of 1869 the dam was finally broken down, the waters poured forth and the lake required about as long to empty as it had required to fill after the original dam was put in. In two or three years Horicon Lake became again Horicon Marsh, a stinking eyesore and depressant to the neighboring population, but perhaps an advantage to duck hunters and trappers of muskrats. Farmers along the Marsh margin were, naturally, able to utilize some land theretofore flowed, particularly for pasture and for hay; but as an offset they suffered in seasons of drought from a lowering of the water table, so that many farmers had to deepen their wells.

Ideas on the subject of conservation were not much developed in the early 1870s; soon after the dam was removed the State sold to a private company the land including the Great Marsh, for about nine cents an acre. The State sold to State Senator Saterlee Clark, the first white man to have visited and commented on the Marsh, and some of his partners, constituting a concern interested in land speculation, 19,850.92 acres for $1,836.37!

The recession of the Horicon Lake waters, completed by 1872, gave the purchasing company an opportunity to parcel out their swamp land. For a time two hunting clubs controlled and monopolized the leases on most of the Marsh. These were organizations of sportsmen who called their clubs the Diana and the Upper Horicon Shooting Clubs. Later on these two clubs joined and became the Diana, but they paid only $200 per year for a complete lease of what Emerson Hough, American author who in 1891 concentrated on wildlife, called "the greatest paradise for wildbirds in America." The Marsh at that time, drained of its great lake waters, was slowly returning to its original state, and the duck population was extremely high. It was estimated that in the late 1870s and early 1880s more than a half million ducks were annually hatched on the Marsh.

The Marsh Is Still A Paradise To Hunters

Friendly Crow At The Marsh

These exclusive shooting clubs, of which wealthy men from Chicago and Milwaukee and New York were for the most part members, require special narration. They were distinctly monopolies and they aroused deep resentment from many local citizens who were not permitted to hunt. On the other hand, the clubs did preserve the character of the Marsh in a most difficult time, and the shooting organizations, to a greater or lesser amount were important on the Marsh until about 1925. The most important clubs besides a very early Caw Caw club, were the Diana, and its counterpart and later joinee, the Upper Horicon Club. But there were other, later clubs, too, that had their own kind of tradition. Strooks Hunting Lodge was a more open and perhaps less selective

group than the Diana, and there were several later others, such as the Redhead Club, Greenhead, that had their own clientele; and several sportsmen's organizations that exist today.

The locations of the clubs are indicated on the "lore" map of the Marsh, but of their presence now there is almost nothing left . . . a few scraps of rock, a piece of iron here and there on Steamboat Island, . . . or Strooks Point that's about all. But the influence of the presence of the Diana Club especially is still very real.

Emerson Hough, before mentioned, was a very well-known American author, born in 1857 and who died in 1924. He wrote a vast amount for nearly all publications having to do with outdoor life and field sports. He was also, of course, the author of the famous books *The Covered Wagon* and *North of 36.* Not so many people know that Hough visited the Horicon Marsh in September of 1891 and attended the opening days of the Diana Club that year. His account, published in the now defunct *Forest and Stream Magazine** lends great authenticity and reality to the presence of the Club and its influence. His references to Marsh history, while they may partially repeat some things already referred to, will merely reinforce an understanding of a complicated subject. Several repetitions may be necessary to make it all clear.

Near the quiet little country village of Horicon, wrote Hough, one of those old, old towns that never grew, you may see today (1891), on the pleasant wooded bluffs which skirt the Great Marsh, the corn hills still intact, which mark the planting grounds of the old Winnebago tribe, the first hungers, probably, who ever tasted the inimitable possibilities of that spot. Probably the Winnebagos did little damage to the fowl beyond stealing an easy livelihood out of their eggs. They were lazy, so lazy that they planted their corn in these same hills, year after year, until the stalks dwindled and dwindled and hardly bore an ear. Even so the tribe dwindled, and when the Diana Club was formed the last of the few Winnebago families around the Great Marsh faded away from the old haunts and silently joined that infinitely pitiful procession of an almost forgotten race.

In the earlier days white hunter and Indian hunter crossed each other's trail in all this country, and the smoke of the wigwam answered that from the cabin of the pioneer. In those days the great strip of rice and boglands was known to the latter as the "Winnebago Marsh." It was then as it is now, between twelve and fifteen miles in length and six miles wide at its

*Founded in 1870 and merged with **Field and Stream** in 1929.

greatest cross section. Then, also, as now, it was a wilderness almost impenetrable of bayou, bog, and slough, covered with the densest conceivable growth of rice, flag, cane, grass, reed, weed, and all the infinite vegetation of the Marsh. The banks were wide apart and gradually sloping. Somewhere between the two rolling banks the channel of the Rock River formulated itself. At the village of Horicon it passed over and through a rocky ledge where it was possible to confine its waters with a dam. At this point that ancient citizen, Mr. Martin Rich, built the dam for his big mill.

"Mart" Rich's dam was built in 1845, when this region was just beginning to settle up. It was a good dam. You can see proof of this yet in the timbers thereof which lie in ruins. It held the waters, but it could not stand the law. The farmers all claimed that it stepped on their riparian toes.

Naturally from the configuration of this country this dam on the ledge at Horicon village backed the waters far and wide over the shallow basin, through which they flowed so sluggishly. The Winnebago Marsh disappeared from the face of the earth, and "Lake Horicon" took its place. This beautiful sheet of water, sixteen by six miles in extent, filled all the basin up and lay high upon the sides of the timbered bluffs, as you may see by the water marks. It was a fine sheet of water. The wildfowl shooting was then simply magnificent, and the fishing also was remarkable. In the winter the ice held hundreds of fishermen's shanties.

Now, all this land under the lake was of questionable value at best, but some men love a lawsuit, and pioneers were ever jealous of their rights. It seemed best for them to sue Martin Rich for backing water over their frog preserves or something of the sort. One by one they sued him, and one by one he beat them, till finally his purse gave out, which in the eye of human justice is the end of everything. So the dam had to come out. Mart Rich destroyed it in the fall of 1867. A great rush of devastating water lasting for days, a destruction of millions of pounds of fish, an unsettling for the nonce of all wildfowl plans, and the great lake was again a great swamp, only this time its name was changed to the "Horicon Marsh," as it is known today.

So the farmers got their hay lands and frog lands back again, and dreamed they owned the earth, because they had purchased or stolen from the Government on very reasonable terms. Speculation in these swamp lands followed, and the upshot of it all was that the taxes never got paid, and the land reverted to the State Government.

Enter now upon the scene one of those things known as a syndicate, in which Senator Saterlee

Sentinel Watches Over The Rookery On Four-Mile Island

Clark, Mr. Trueman A. Woodford, Mr. Shepherd, and Mr. G. H. Lawrence figured principally. This syndicate was known as the "Mechanics' Union Manufacturing Company," and it bought of the Government at auction sale at a price as low as seven cents an acre a large body of the 25,000 acres covered by the Marsh.

The wildfowl clung steadily to the Lake or Marsh all this time, paying no attention to the litigation, and within the last decade the desirability of a great wildfowl preserve here became apparent to a number of Wisconsin sportsmen. The Diana Shooting Club of Horicon was formed and incorporated. The charter members being Messrs. S. E. Davis, W. A. Van Brunt, C. A. Hart, J. B. Hayes and Geo. H. Lawrence. June 8, 1883, was the date of the club's incorporation, but on Marsh 20, 1888, a lease had been arranged with the syndicate above mentioned, by which 5,800 acres of land, comprising the southern and best portion of the Marsh, had been leased for twenty-five years, the annual rental being $100. This lease has now been provided for out of the club's sinking fund.

Originally only twenty-five memberships were sold, and the bulk of these were placed in Horicon and Milwaukee at first. It chanced that Mr. Percy F. Stone, one of our prominent Chicago shooters, who was born and raised in Wisconsin, happened to be passing through that part of the State when he heard of the starting of this club, in the spring of '83. He knew the Marsh, and was interested at once, so much so that he telegraphed to the incorporators to meet him at the train as he passed south. To them he said, in brief, and while the train was waiting, "I will take five memberships." Arriving at Chicago, he forthwith took in on his spare tickets his friends J. O. Pierson, C. E. Cole, E. J. Marsh and C. C. Germain. These five formed a club within a club, and put up what is known as the Chicago Shooting Box, where things were long run in a highly exclusive and bang-up style, dinner in courses, white caps on the cooks and waiters, and all that sort of thing. This comfortable club house, four rooms, with plenty of bunks and lockers for chance guests, made the only club building on the grounds. A Milwaukee and Horicon contingent built a big and roomy cabin boat, which is annually anchored (1891) at the edge of the Marsh, as high up as the channel will permit, and other parties have put up shanties on the drier portions of the Marsh, as may be seen by reference to the accompanying map. The latter is the personal work of Mr. Stone, and it is well done and highly useful, both to the reader and to shooters interested in this Marsh. As explanatory of the map, we may quote from the club history written some years ago by a gentleman who was a member of Congress, or of the Legislature, or something, and who may therefore be supposed to be way off better able to write about it than I am: Nearly every species of migratory waterfowl known to the inland waters of North America can be found here. Mallard and blue-winged teal breed here in immense numbers. Wood duck, widgeon, spoonbill and redhead are plentiful. The canvasback is occasionally taken, and when the autumn flight southward fairly sets in, the pintails seem to outnumber all other varieties.

While this entire Marsh tract presents attractions for sportsmen unsurpassed, except in the far West and North, the grounds controlled by this club are the choicest to be found in the extensive area covered by the Marsh. They are easy of access, especially from Horicon by the river, which winds circuitously northward for a distance of about three miles, when it becomes lost in an infinite number of channels, bayous and sloughs trending in every direction and of every conceivable shape. Following to the eastward this labyrinth of bog and channel, one soon reaches the "Four-Mile-Island Bay," a sheet of water covering an area of about three hundred acres. On its eastern shore rises a ridge of hard land, not many years ago denuded of its timber, but now thickly covered with a growth of young trees, which is known as "Four-Mile-Island," the favorite resort of camping parties. To the north of this bay is the "Skirmish Line." A channel connects the Four-Mile-Island Bay with a sheet of water covering about an equal area called "Malzahn's Bay," into which the east branch of the river debouches its waters and affords access to the grounds from the east. West of the Four-Mile-Island Bay, and connected with it by several channels, is a sheet of water covering about one hundred acres, which from the shallowness of the water and the interminable depth of the mud, is known as "Mud Bay." West of this, one again enters the maze of bog and bayou, but after winding a distance of about two miles, reaches another sheet of water of about two hundred acres called "Mieske's Bay." A mile north of this is the largest sheet of water on the grounds of the club, known as the "Big Lake," and covering about twelve hundred acres, and continuously supplied by the waters of the West Branch. From all these principal sheets of water innumerable inlets, estuaries, bayous and sloughs, bordered with sedges, flags, reeds and the various rank vegetation of the Marsh, trend and wind in all directions and for seemingly endless distances, thus forming thousands upon thousands of acres fitted to the very highest degree by nature as the home of the waterfowl.

Once lost in the infinite intricacies of these morasses, at the gloaming just preceding nightfall, when the myriad animal life is momentarily silent, one cannot fail to be deeply impressed with the sense of utter

seclusion and desolation which everywhere pervades the place. But in a hazy autumn afternoon, when the fields and woodlands that fringe the great basin in which you stand are clothed in brown, and yellow, and crimson, and gold all around the horizon, and the western sky is bathed in the warm, soft tints of an October sunset, the desolate, uninviting picture, upon which a mere speck, you, are penciled, is lost and forgotten in the gorgeous beauty of the frame in which it is set.

It will be observed that the grounds leased by the Diana or "Horicon" Club are about five by three miles in extent, and cover all the open waters and channels of the Marsh. Another club, with headquarters now at Milwaukee (The Upper Horicon), controls all the Marsh north of the Diana grounds; so that the whole Marsh is preserved. A visit to the Marsh shows it to be the most easily guarded territory imaginable. The ground covered by the Marsh growth is too soft for a trespasser to attempt on foot, and the few channels which lead into the shooting places can easily be covered by two or three watchmen. The club has two regular watchmen, both deputy wardens. Trespassing is practically unknown. For a time the club sold shooting permits, but even these have now been abolished, and a few guns only own this magnificent preserve. The result of all these natural and artificial conditions is that the supply of wildfowl is simply astonishing, as I shall later mention in detail. At this very point, however, it should be chronicled that, law or no law, the club never has nor will permit spring shooting.

The Chicago Shooting Box pursued its way very happily until death called away two of its members: Mr. Cole and Mr. Pierson . . . Their places were filled, but finally things so shaped themselves that it seemed wiser to turn in this property to the club, to extend the sale of memberships up to forty out of the fifty to which the charter limits the club, to protect the lease, make some improvements and eventually to build a larger club house. In fine, this is the present policy of the club, the forty mark has already practically been reached. The memberships are rapidly taken at the ridiculously low sum of $150, this, with

A Retriever Gets A Workout During One Of Many Field Trials On The Marsh

the annual dues of $15, being cheap pay for such privileges. Little money is needed for spending on the Marsh, for it needs little help to make it perfect, barring the few watchmen and the expense of a ditch or two. The revenue from the dues will soon be enough to cover the expenses of this tight little corporation. Another novel source of revenue is the "rat money." The Marsh swarms with muskrats. Last year the club

Baby Terns Along The Dike

leased the trapping right to "Old Man Mieske," a German farmer whose land adjoins the Marsh, and who has been a great help to the club in many ways. Mieske cleared $400 on the rat deal that year. An odd character, Mieske. He came to this spot forty years ago, when the country was perfectly wild. His fine house and barns show his success, as well as do his broad acres. He bought his farm with wild ducks,

baited and slaughtered with a four bore. He killed ninety-three ducks at one shot, once upon a time. The ducks, the rats, and last of all the club men, who need teams, pushers, milk, eggs, chickens and what not, have in turn yielded golden harvest to this thrifty, dried-up, but still hearty and vigorous old-man-of-the-Marsh.

Connected with the club from the start, and so thoroughly acquainted with its requirements, it was natural that Mr. Percy Stone, so good an executive officer withal, should have a prominent place in all this recent work of reorganization and preparation. It was Mr. Stone that shouldered the responsibilities and infinite detail of getting ready for the opening of the season this fall, and converged all the chaotic wishes of his fellow members into the happy focus of that event so eagerly awaited each year, the day of days, the "opening day." This year he had an additional and most serious source of trouble: the water on the Marsh was too low to let in the boats. This occurs about once or twice in five or six years, and while it does not affect the number of ducks, it does seriously affect the shooting. About four weeks ago, therefore, Mr. Stone had a dam built across the river, just below the "first channel," about opposite the spot marked on the map as "Stony Island." The river here runs through banks of peat and muck, its bottom yards down in fairly fathomless mud. The dam was built of plank, peat, hay, and mud. The river promptly burrowed under the dam, and blew a great hole out of it. This was temporarily filled, and nineteen inches of water at the dam set the water back in the channels so that the boats could go in all right. This was just in the last of August, but unfortunately, only a few days before the momentous first, Mr. Stone got a telegram that the dam had gone out again. It was therefore with some despondency that the shooting party started for the Marsh. Work on the dam was continued persistently, and on Sunday, the thirtieth, the leak was gotten nearly under control. There seemed to be no bottom to the mud. I saw a sixteen-foot fence board driven clear to the end into the mud at the dam, and the last blows of the sledge seemed to send it on down as easily as the first. No one can fathom the mud on Horicon Marsh. When this was Horicon Lake, the whole bog, rice, reeds and all, rose and floated, miles and miles of it. When the water was drained off most of the bog sank again. If the water rises the bog rises also. There are two kinds of mud on Horicon, the kind that floats and the kind that sinks. If you break through the bog while wading through the wild rice, you may sink up to your hips, or up to your waist. You may strike terra firma in a layer of sunken peat, or in one of the solid ledges which underlie portions of the Marsh; or then again you may sink a thousand feet into the soft mud, the kind which doesn't float. I dropped my jack knife while putting out my decoys. It fell in three inches of water, but though I plunged my arm up to the shoulder after it, I never caught it, and think it is going yet. Study the infinite combinations of these two kinds of mud, the kind that floats and the kind that does not float, and you will soon learn that pushing a boat on Horicon Marsh is better when the dam holds.

It held all through the first, second and third, but the water rose all too slowly. This dam cannot be permanent. It is too much an engineering feat to make one permanent on that side of Horicon Ledge; but granted that, and you have the key to the prettiest shooting situation in this part of the West. We have no marsh which approaches it. The fastnesses of the Marsh protect the birds, and they are there in untold thousands.

Marsh Newcomers

In view of what Emerson Hough had written in 1891 it was fascinating to me, to talk personally in the spring of 1972 with Hubert Pagel, eighty-two, small, wiry, extremely bright and dynamic, long a baker in Watertown, Wisconsin. Hubert was born on the Marsh, at Horicon, in 1890, one of twelve children. In his young manhood he knew the Marsh as well as any person, for his father was caretaker for the Diana Club and his mother was the cook for the hunters when they came each early fall. Hubert now lives above the bakery in Watertown, retired, and has a fine collection of Indian relics, mostly picked up on the Marsh.

The Diana Lodge, Hubert told me, was a beautiful building. It was all made of Georgia pine and inside, the great beams and large open space seemed unbelievably spacious. Around the main lodge were a lot of individual cabins where the various members stayed when they came to shoot. There were probably at least three dozen "shacks" and some of them weren't so small, either. The hunting season opened first of September, and my mother was the cook of the Diana Lodge and my father was the caretaker. He got thirty dollars a month, for looking after the Marsh. The club had it all leased. Our family could trap all the muskrats we wanted and we used to shoot ducks for them hunters, and we'd have ducks hangin' in the boathouse, twenty-five to the bunch. The hunters would come out to the boathouse and pick up the ducks and take 'em in to Milwaukee . . . ducks we shot for 'em. The lodge had a great fireplace too. Each shack would have its own boathouse, and from their individual landings the hunters could push out into the Marsh. Us boys done the pushin'.

Mother gave the club members regular home cookin'. Among the other good things she cooked sometimes was muskrats. We would take the rats and soak 'em good in water, and she would cook them up, and the Diana Club members would eat 'em like greased lightning! Boy, was we really eatin' muskrats them days!

We'd go out and shoot them big prairie chicken, too, and someday I'm gonna ask the "conservation" (now Department of Natural Resources) why they didn't put them big prairie chicken back in the Marsh, ain't any there now. Those years back around the turn of the century, the Marsh was loaded with prairie chicken. The club members ate mostly game. They ate ducks, and fish and prairie chicken . . . we used to go down to the Mieske Ditch and get a fifty-pound sack of perch in an hour! They would weigh a

Hubert Pagel, Son Of The Diana Hunting Club Caretaker Remembers . . .

pound apiece! And when my brother took my father's place as caretaker, after my dad died about 1904, they would get sometimes five hundred pounds of frog legs. Above the Diana Club dam, you could go and look through that water, and it was just black in there with frogs. You could take a landin' net and scoop 'em out; they cleaned them frogs and used to ship 'em down to Milwaukee. Flooded the market down in Milwaukee with Horicon Marsh frog legs. Got so they wouldn't take 'em in Milwaukee. All the Milwaukee bellies was full of Horicon frog!

And along with the game, which was the staple food at the Diana Lodge, my mother would fix potatoes, hot biscuits, coffee cake, homemade bread, and gravy such as you never tasted. She'd bring out a whole great pan of teal, jusk baked brown and crisp! That was one reason the Diana Club members liked to come . . . the home-cooked meals. We had three cows and would drive 'em up to the Diana in the early fall, and mother would make butter and cheese. Her butter was the best I have ever eaten, all home churned with a big dasher churn that she chugged up and down in a big crockery jar.

There never were any women come to the Diana. All men and only men. Didn't want the women there. Spoilt the fun. Guess they went to town when they wanted women.

I was a great hunter, though I was only about sixteen years old. I had me an old ten-gauge greener, a double-barrel shotgun with exceptional long barrels . . . goose guns they sometimes called 'em. None of the Diana Club men had automatic shotguns in them days. Most of 'em had single-barreled or double-barreled Ithaca guns.

I was as well acquainted with the Marsh in them days as anybody. My brother and I would go up into the Marsh and hook the wild rice out of the ditches so the hunters could get their boats through. And we got fifty cents an hour for pushing the hunters' boats . . . long poles that we pushed with, keeping very still, so we wouldn't scare the birds off. We push through from the Diana Shootin' Club over to Mieske's Bay and then to West Lake, this side of Steamboat Island over to Four-Mile Pond, and we had to clean them ditches all the way along there. In those days the ditches was loaded with wild rice, everything was wild rice. It was so common, and we fought it so to help the hunters get through, that we never ate any of it with meals. Mother never cooked any wild rice. Was thought to be just a weed.

I remember once a flock of about three hundred teal come in, comin' in to land, hard, you know, and with one shot I killed fifteen of them teal . . . They was comin' in to eat the wild rice.

But we had some wonderful fishin', too. I remember how we'd go up to the Diana Dam every day. We'd go up with a big rake and when we had the wild rice out of the way so the hunters could come through, we'd stop at the dam and we'd always fill up sacks with big fish. We'd bring 'em home, and Mother would pickle 'em, or give 'em to the neighbors . . . pickled and smoked and everything else.

We'd catch the fish with our hands. There was that runway where you went across with the boats, and then it would open out above the dam by the willows. The fish would go in between the dam . . . in that runway, and we'd go in there and catch pike . . . northerns weighing maybe twenty-five pounds!

I remember old Governor Philipp, Governor of Wisconsin (1915-1921), he'd come out to the Lodge to hunt. And all the members would congregate after supper in the big Lodge which was just fitted up for a club room . . . no meals were served there . . . Mother had a special kitchen and dining room built separate. And after a day of huntin' the members would be tired and there was always plenty to drink. Some of the big brewery men, Pabst and them fellers in Milwaukee was members and they would keep the Diana well supplied with beer and liquor. There would be a big fire in the fireplace and one of the evening sports was to throw live shotgun shells into the fireplace and let 'em blow up, and the shot would roll all over the floor! Old Governor Philipp especially he would get a kick out of that! He would beller with laughter and toss in another handful of shells. Some fancy dancin' then took place, but nobody ever got hurt. I remember when they first came up to the Diana in the first cars. There was an old dirt road runnin' down to the main road . . . cars would get stuck in there . . . Fords, Cadillacs, Packards, Stutzes, Kissels, every car on that mud road! Stuck clean to the runnin' board. Swear and cuss! and we'd have to help 'em out.

Most Prefer The Top Wire

Duck Hatches Were Important In The Old Days, Too.

Early Bird Life On The Marsh

It wasn't so easy to keep the ordinary citizen off of the shooting club grounds. Walter Frautschi, a sportsman, Madison, Wisconsin, tells how a disagreeable situation arose from the action of the Lombard Investment Company — one of many promotional groups anxious to exploit the Marsh — which, disregarding the prior leases of the combined Diana and Horicon Clubs, gave game preserve rights to W. R. Grady of Chicago. He established various agencies throughout the country and sold permits to hunt on Horicon Marsh at $3 each. When the hunting season opened on September 1, 1894, the preserve was swarming with those permit holders, most of them the innocent victims of Grady, but others wilful trespassers. At the instigation of the clubs, deputy sheriffs and marshals were sworn and in a short time the Dodge County court dockets were filled with names of persons they caught. Forced to take action, Grady asked for an injunction to establish the priority of his claims with the result that the clubs won a complete victory and a sympathetic nod from the judge.

Life at Horicon was not all aggravating worry about defense of the property, nor was it all bonanza shooting. There was much fun; there were occasional instances where man's frailties transcended sportsmanship. Sometimes there was a near tragedy, as when R. M. Rogers becoming lost on the Marsh without a partner, almost perished during a lonely night in a November blizzard. Showing great presence of mind he saved himself by buttoning up two live decoys underneath his hunting jacket where their body warmth contributed to his salvation. By and large, however, the quarter of a century after 1883 witnessed a procession of sporting friends and outdoor folks whose major interest was shooting, but whose capacity for companionship superseded all other considerations.

Frank Bodden, great old-timer of Horicon, and an original member of the Diana Club told Barney Wanie that the club members weren't particular about what kind of ducks they shot. They would shoot at anything that came along, he told Wanie in 1968, when Frank was ninety-eight years old. At the beginning of the season was when the teal would come. Thousands, and thousands and thousands of 'em, as Frank said. The teal were so thick that when they took up off the water it roared like a thunderstorm, you know, and you couldn't look through 'em any more than you can through some of these dense swarms of blackbirds that you see now in the Marsh sometimes. But when the teal left, then the mallards would come. Mieske's Bay was the place where they used to go to shoot teal. The big ducks seemed to come most on the larger waters: West Lake, Lewis' Hole, West End,

and Four-Mile, they were all big ducks mostly. The teal, though, gathered at the small hole right at the Diana Dam. And what sport it was, said old Frank, to draw down the pintails. And John Yorgey was a master at calling. If there was a flight of thousands, it seemed, and John would start his calling, the birds would begin to circle and circle and down, down and right to it. The club set a limit of twenty-five ducks a day, but before 1885 or so there wasn't any limit and some of the fellows just competed with one another to see how many mallards they could kill in a day. For a long while the record was held by a hunter named Melcher who killed a hundred and sixty-seven mallards in one day; that was the envy of John Yorgey and there was always a rivalry between those two fellows who could kill the most mallards in a day. And finally John Yorgey beat Rogers and shot one hundred and seventy mallards in one day. That was the record. At that time they could ship these ducks to a market in the city, and it wasn't entirely just a waste of ducks. At least it was then considered legal if not ethical.

Way, way late in the fall, Frank said, there might be one flock of wild geese. But geese were really rare on the Marsh. Frank never killed a goose. But there were some flights of swans into the Marsh in 1900, and Frank saw them come in; but they never stayed very long. Just in and out. Always there were cranes. Frank didn't remember any sandhill cranes, though he thought there might well have been some. Nobody made a distinction in those days, that was all. George Hall, he said, was certainly one of the very first wardens on the Marsh, and Frank recalled that George must have come in there about 1925, though he wasn't really sure what year. The old Diana Club wouldn't allow any "baiting" of the birds to draw them down . . . but the old commercial hunters did that. Frank said that the members of the Diana Club really did practise a kind of conservation. The attitude was just different, then, that was all. There just seemed to them so many ducks that they could never possibly be all killed. Nevertheless the club members did what they could to protect the fowl. They did use decoys, wooden, and hand-carved for the most part, but they allowed no live decoys. Even now some of these old club wooden decoys can be seen around the Marsh, and I found one not very long ago in a farmer's yard at LeRoy. It was just sitting there by the doorstep, very lonely and forlorn, completely out of its time.

And of course as time went on other clubs came. A big hunting lodge, on the east side of the Marsh, and somewhat later, was quite a place too. One old-timer with whom I spoke, said:

In them days they had that big, well, it was like a hotel down there on the east side. Hunters come from all over . . . Chicago, Milwaukee, all around. Hunted and stayed there. Could eat, sleep and drink there. 'Specially drink. They had guys that used to shoot ducks for these fellows and them days they shot barrels of ducks. They sold 'em to these hunters that come in from different places and was too lazy to hunt themselves, so they just bought ducks to take home. Lot of the guys didn't hunt ever. Got keyed up drunk and one thing and another. They sometimes had girls there, and they had one hell of a time. I guess they had a kind of limit of twenty-one ducks a day, but not many observed that. There wasn't only one game warden on the whole damn Marsh.

Barney Wanie Above The State Dam In Horicon

THE COMMERCIAL HUNTERS

When I was visiting with Barney Wanie who worked on rough fish control on the Marsh for close to forty years, he was telling me that the Marsh once had some commercial hunting. In the older days in the 1880s for example, said Barney, sure some guys come in here to kill the ducks particularly. Wasn't any geese, then. And some of 'em had these big bore guns . . . four-gauge, if you can imagine what a big hole a four-gauge gun has in the barrel! And they would pour a double handful of black powder down the barrel, and a few pounds of shot, and ram down some wadding, and put a percussion cap on the nipple, and set there waiting for the birds.

The old gun was always on a rest, or a stand, sometimes, because she was so big and heavy and kicked so bad that a single man couldn't hold her. And when they heard the flight comin' they leaned back out of sight and eared back the hammer . . . pretty big and heavy, and the spring real strong. Had to pull her back hard. Then come the ducks! What a sight, about half a mile of them, and they come in and circled once or twice and set to come down. Then the men got ready, and when the ducks came in, and set on the Marsh, so thick there wasn't room between them, they aimed the Marsh gun or "Punt gun" down through the middle of 'em, and got her set, and the guy who was back there on the trigger, he set his eye down the barrel and figured he was in line and maybe he shut his eyes, because that old gun made one hell of a noise, I dunno. Anyhow he pulls the trigger and a spurt of fire about fifteen or twenty feet long leaps out of the barrel and the damndest bang you ever heard and all that bird shot smoked out of the barrel and struck amongst the ducks. A terrible slaughter it was . . . sometimes as many as a hundred ducks would be killed with one shot of that old betsy. Illegal now of course. But not then . . .

There was a farmer lived on the Marsh in the old days back in last century. And this farmer was named Mieske. He had a green thumb. He could grow anything. And his relatives live in Horicon right today. They got green thumbs, too. They can grow stuff real good. Another fellow had a stone pile out there on his farm. He was one who had one of them big guns. A four-gauge he had . . . He would load that thing up with shot . . . and he would feed the ducks for a few days until they came in real good and got real tame. Then he would let the old swamp gun go right into 'em. Would pick up so many birds he never rightly knew how many he killed. Then he would put the ducks in barrels, on ice, and ship 'em to Chicago and Milwaukee just as they were, not picked or anything,

just the innards taken out . . . Yep, there was some who did that. They would cut the ice in the river, of course, when it was good and thick in the winter, and store it up in sawdust piles, or in an ice house that would keep it indefinitely. And they shipped those ducks to Chicago, and the restaurants there would pay good prices for them because the Horicon ducks was all fat, like as not. Good feeding here, because when the fire would burn down potholes, and burn the peat out, and then new vegetation would start up, it made real good duck feed. They liked to come here.

Well, the Diana Shooting Club didn't like this market hunting . . . this commercial hunting. It ruined their sport and did big damage to the duck flocks. So they had a guy who was a watchman for 'em. You heard of Bismarck, haven't you? The German chancellor back last century? Well, this watchman was a big German Bismarck kind of guy, and he stood around with his chest out. And they called him Old Bismarck.* He was the deputy, see? And he was tough, I can tell you. He caught a guy on the Marsh with one of them big cannons, he would throw him right in, and his big gun too if he could get hold of it. Old Bismarck.

And you might think that the commercial hunters would have a field day with all the wild geese there were supposed to be here then, too. Because if you could get a hundred ducks with one shot you might easy get twice that many geese, them bigger and all. But I'm here to tell you something. There just weren't any wild geese here those days . . . not until 1941 did I remember I saw any geese to amount to anything, 1941 it was, I remember, and that year Delmer Boyer killed a wild goose on the Marsh.

The geese came, later of course, because of the Refuge. Some geese stopped and the Marsh men got a few of 'em, to clip their wings to stay; and that few came back and came back.

Barney Wanie told me that when the Diana Shooting Club was organized, the club members (and this furthers Hubert Pagel's story) had a channel dug from the high shore near the club to the main channel of the Rock River. This was called the Diana Ditch. Barney said (as did Hough) that they put also a small dam in the same area where the ditch was dug, and they put some planks in endwise and that small dam backed up the water about a foot . . . just some boards with a little mud against them. This created a little wider area on the river, and that area, when the Diana Club members hunted there in the 1880s and '90s, was so thick with wild rice that they hired farm-

*Bismarck was Hubert Pagel's father.

Carp Fishermen

ers to go in with boats and cut the rice out, so there would be room for the club members to shoot, and to put their decoys in the water.

Now, Barney says, I take a lot of visitors out into the Marsh, and I can show them, along the channel, at one place, three spears of wild rice. That's about all that is left of that vast harvest that was right there in that place in prehistoric times, and when the Indians we know came, and when the first white men came, the wild rice was so thick you couldn't walk through it. Now about three spears only are left. This rice is growing on a piece of higher ground about a thousand feet from the old Diana Dam . . . or where it used to be. I know where there are a few other spears, too, in different places, but in this place where the rice was unbelievably thick . . . only three spears.

Wild rice, Barney continued, contrary to what a lot of folks think, grows on fairly solid ground. It doesn't do well on loose mud. So after the Marsh was drained down, as they did, you know, in the early years of this century, and then was reflooded, a lot of the silt was washed in. The wild rice wouldn't grow in that silt. It just quit. Died out. There was a profusion of wild rice in the east branch of the Rock River, too, just a little east of Four-Mile Island, where the heron rookery is now. The bottom of the river there was all solid, and there are still a few stands of wild rice in there. Unfortunately the main river is closed off by a floating bog . . . this happened when the Marsh was reflooded, and the whole bottom of the bog just rose to the surface and started to float. That bog is about a quarter of a mile wide and a quarter of a mile long. It goes to show how the whole character of the Marsh has been changed by the draining and flooding that has been done. Once I attached a launch to a great floating bog and tried to move it . . . it was over a thousand feet long. I couldn't get her going!

Rough Fish Control

Clyde Huffman's Family On The Dredge

The old Indians in their teepees in the skies must be very puzzled over the white man's antics in regard to their sacred Marsh. Doubtless, if they could, they would certainly ask the Great Spirit to remove all their ceremonial mounds from a place cursed by the greed of men who were so willing to bring ruin to the richest storehouse of plenty that the red man ever knew. Nature, for all her inconsistancies of famine and plenty, of heat, and cold, and water, brought things out all right in the long run. The wildlife persisted, and man, so long as he was the child of nature, persisted too.

As one old German farmer, Bill Schreiber, told me, the Marsh became just a playground for the lawyers. And certainly, the wetlands have been in the courts more than any other single tract of land in the history of Wisconsin. The great heyday of the shooting clubs was probably about 1900 and then in 1904 a grand plan was hatched to make the Marsh into an area of fine, drained, agricultural garden. The thing was engineered by a Chicago banker, W. C. Norris, and by Illinois developers who bought about 18,000 acres in the center of the Marsh. And the farmers around the edges became intrigued because they could see how their own enterprises would benefit . . . more farmland, better growing conditions. No more flooding of their Marsh-edge farmlands in a wet year.

The drainage ditch plan was quite simple. It consisted of a main ditch for the entire length of the Marsh, with laterals at one mile intervals, nine laterals on the west side, eight on the east. This main ditch was to be dug along the range line which divides the Marsh through the center. Its effect was to straighten and lower the river channel, thus increasing the speed of flow through the Marsh, and to lower the water table. This effect was precisely as intended, but unfortunately it did not eventually make the Marsh suitable for agriculture, though the owners of bordering lands were able to cultivate their lands more effectively. The principal effect was to impair the Marsh for its original vegetation, to make it more favorable to the growth of weeds, and to increase the danger of fires, of which several bad ones actually occurred.

The drainage scheme was opposed by many sportsmen, and by others up and down the Rock River who foresaw dangerous flood conditions if the dam at Horicon and the dam at Hustisford down river were to be removed, and they would have to be removed in order to drain the Marsh. Of course the matter reached the courts and on April 17, 1908, the Wisconsin Supreme Court handed down a decision that Rock River was a navigable stream, and that no authority of law was delegated to impair or appropriate it to drainage purposes, and that the drainage proposed would have that effect.

It would have seemed that the drainage men, who had already invested several hundred thousand dollars, were defeated. Not so.

The drainage interests went right ahead, even in the face of the Supreme Court ruling. The reason they were able to do this was probably because of a split-court feeling that Rock River was not *really* a navigable stream, and the fact that it was navigable on paper should not allow a court decision to block a plan that might bring great economic good to the whole area and the State.

And too, the drainage plan did have the support of a lot of people, including the people at Horicon who eventually allowed the local bridge and the dam to be removed by the drainers. The farmers, for the most part, could see only good in the program.

Having heard something of this old controversy, I went one cold, February morning to Waupun to see Clyde Huffman. His father, I had been told, was the engineer who supervised the dredging of the Marsh . . . back in the early days of this century. I had been

Clyde Huffman, Marsh Authority

told so many things about the damage the dredge did to the Marsh, and to its ecology that I wanted a first-hand account of just what did happen and how.

I discovered that Clyde had an apartment above Brown's Cafe on the Waupun Main Street. I went up a very slippery wooden stair at the back of the building and knocked at Clyde's door. Nobody was at home, but in the window was a sign that said Post Office and another one that said Food Store. I gathered that Clyde might be at one of these spots and so set out to find him. I never did find him at the post office or the store, but at each place word seeped to Clyde wherever it was that he was hiding, that a stranger carrying a strange-looking "carpet bag" wanted to see him. The bag, incidently, was my tape recorder.

Eventually, I met an old fellow in a brown wool jacket and wearing a thick wool hat and asked if he had seen Clyde Huffman. He looked at me carefully as though I might be crazy and said, "What do you want with Clyde?"

"I wanted to ask him about his father."

"He have a father?"

"So I been told."

"I hope he did," the old man said, "because if he didn't then you wouldn't be standin' here talkin' to Clyde Huffman!"

We went over to Clyde's place, and I had to pause inside the door in sheer amazed admiration. Clyde has saved every newspaper for the past forty years, and he has them all, or parts anyhow, stacked and piled up in his apartment. There are boxes all filled with clippings, too, and all lettered on the outside to show what each box contains.

Clyde draws out an old rocking chair and I sit in one of his kitchen chairs. He is a man you will never forget, once you have seen him. He must have been an exceptionally powerful man in his time. His shoulders are still massive. He talks slowly, very clearly, and you certainly don't need an earphone to hear him either.

My father worked in those days, before 1910, said Clyde, for the G. A. MacWilliams Drainage Company. They sent him up to the Marsh to operate a dredge. They sent the parts of the dredge by rail from where they were made in Indiana, and the whole dredge was put together at the north end of the Marsh . . . what they called Dennis Conners Point.

While the dredge was being constructed, Clyde's father, John S. Huffman, found a house to which he could bring his family. The house he found for them was at Chester, on the west side of the Marsh. It was a very small place, with only a few houses.

The dredge they built was about seventy feet long,

and it would dig out a huge, dripping mawfull of earth at one bite. The dredge was operated by steam and was eighteen feet wide. It was a floating dredge.

The way they got the dredge into operation was by digging a big hole around a spring marsh, a wet place that would fill with water. They built the dredge right in the hole. Then when the hole filled and floated the dredge, they just started to steam shovel their way on out.

Hawks Also Monitor The Marsh

John had been at the Marsh for a couple of months before the family arrived. There was John's wife, and two boys and a little girl. They set up housekeeping in a large frame house at Chester, on the main corner.

The dredge dug the side ditches about twenty-four feet wide, and it had "spud" arms on the sides to hold the dredge solidly in place while it was digging. When the dredge had dug as far ahead as the long shovel arm would go, then they had to float the dredge on ahead, and reset the support arms. They dug straight south from the spring hole at the west side of the Marsh.

Eventually, by side ditching, they joined the main channel of the Rock River, or used the water from the Rock River to float the dredge. It was almost fifteen miles from the first hole that the dredge made, to the bridge down at Horicon.

The purpose of digging the main ditch and side

Early Spring Swans And Geese On The Marsh

Deer Play On The Marsh In Spring

104

ditches was, of course, to drain the Marsh. The whole idea was that the Marsh would be reshaped into a garden. The developers had selected a name already. They were going to call the Marsh "The Garden Spot of the World." They published a large bulletin, *Onions and Independence,* which described the Marsh, and its advantages as a vegetable growing location, and the book pictured, falsely, scenes which were supposed to be taken at the Marsh, but which were really taken elsewhere. The largest onions in the world were to be grown on the Marsh, and all other vegetables of like size. People were to live in utter plentitude, and achieve lives of such wealth and splendor that they would be transported to a veritable heaven, entirely created through the draining of the Horicon Marsh. Marsh land was sold to persons who had the get-rich fever. It looked as though, for a time, that the Marsh agricultural project of growing excellent vegetables was going to succeed. I am jumping ahead a little, now, but after the main ditch had been dug, and things tidied up, about 1916, the promoters put an "experimental farm" of about 300 acres at the north end of the Marsh, near where there is a Federal "goose check" station now. They planted this plot to onions, celery, potatoes, rye, wheat, barley . . . anything to make a show. And they were fortunate in having just the right amount of rain in the first season, 1916. The fields were perfect. Rows of onions nearly a mile long, perfectly straight, and of lush size. The sight of the beautiful experimental farm led to hundreds of purchases. People were paying $500 an acre. Then the scheme blew. Next year a deluge of rain fell. The whole experimental farm was covered with water. The drainage project never bounced back from that.

As the ditch was dug, John Huffman, the engineer, grew more and more doubtful about its success. There was simply too much water in the Marsh, and he believed, rightly that no ditching operation would ever clean it out enough to make the land entirely tillable.

John got about $150 a month for his work as dredge engineer. He could keep his family very well on that money in those days, and he greatly enjoyed the work. There were thousands of ducks, and each morning the ducks would fly up when the work began. John felt that he was in the very middle of nature. There were few geese, however.

Clyde was with his father as much as he could possibly manage it. He, too, loved the Marsh and grew to love the sound of the dredge, and the sight of the great shovel eating away the earth and depositing it at the sides of the great ditch.

Clyde told me that the entire crew on the dredge was seven men besides John. When the ditch got

about opposite the place where the old Strooks Shooting Club had its headquarters, a group of hunters came running out to the dredge, threatening the crew with shotguns. The hunters, who were fearful for the preservation of their hunting grounds, since the dredge would probably drain the wetlands where the ducks nested, dared John Huffman to continue. They would kill him and destroy his dredge, and shoot every man on the dredge if it continued to dig.

Clyde doesn't remember his father's exact words, but in effect he said his father told the hunters that he wouldn't stop. He was hired to do a job, and right or wrong, he would continue. He stood dauntlessly in the bow of the dredge like an admiral under fire, and dared the hunters to start shootin'! He simply started the dredge digging again, and the hunters stood by futilely.

The men on the dredge didn't care much for wild duck, as far as that was concerned. The Huffmans had come from a farming area in Illinois, and they weren't used to wild game. Ducks had for the Huffmans then a sort of rank, wild taste that they couldn't get used to, and they weren't very expert at cooking the game to reduce the wildness. Wild ducks in those days were being killed on the Marsh by the thousands, and many were still being shipped in barrels as far away as New York; but very few ducks were given to the dredge men. The hunters hated them so much, and many local people too were very suspicious. The dredgers sometimes went hunting themselves.

On the dredge crew was a cook, and sometimes it would be a husband and wife team, the wife cooking and the man working on the dredge. Meals were very

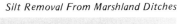

Silt Removal From Marshland Ditches

hearty. Beef, pork, potatoes, plenty of thick gravy, hot biscuits, honey, sorghum, country butter, thick fruit pies . . . these were the standard items of the dinners.

In 1917 the Huffmans moved from Chester to a house on Stony Island in the Marsh. The waters surrounded the island completely at high water time in the spring and fall, but other times one could walk to the island. Clyde thinks that it was the walks to the island that opened his eyes more than anything else to the life of the Marsh. He remembers the hawks that circled so high, and came down, occasionally, in a hurtling dive on a bird or a rabbit; he recalls the cries of the heron, and the heavy lift of the egret, and the way the blackbirds came across the Marsh sometimes in a cloud a half mile long.

The dredge drove and tore right down the middle of the Marsh and into the city of Horicon. The bridges were all removed so the dredge could go through, deepening the channel of the Rock River. The glacial dike across the river, which had furnished the initial thoughts of men like Mart Rich about damming the river for water power, was dug out. Young Clyde Huffman, in fact, was the one who set the dynamite charges that blew out the old dam . . . same place where the first dam had been built in 1846. He blew that old dam in 1914 when he was a member of his father's dredge crew.

Harbingers Of Spring

106

Fire On The Marsh Near Horicon — Marsh Fires Have Both Value And Danger

THE HAY MAKERS

Draining the Marsh in the early years of this century caused a disaster. The disappearance of the waters left, after a while, a tangle of weeds and low brush. The wildlife, birds and geese, departed. The wild foods, rice and cranberries, which had drawn the Indians to the Marsh for centuries died out. Fires started in the tangle and in the peat. Smokes rose from the Marsh which could be seen for almost a hundred miles, and thick ashes covered the whole surrounding country.

The men who made hay around the Marsh were especially conscious of the fires. The fires destroyed the standing marsh grass, and were also a terrible menace to the stacks of hay put up by the harvesters.

Beginning with 1880 and extending down to the time of the drainage operations, landowners, even with the several dams then in existence, were able to cut large quantities of hay off the Horicon Marsh. This was true even of the lands covered by the Diana leases. The land was in such shape that during the hunting season prairie chickens could be hunted everywhere on the Marsh uplands. During these drainage years there was small water on most of the haylands.

Eddie Lechner and I went to see Pete Feucht, son of Marsh pioneers, whose father had the first hay baler on the Marsh, and whose very life was the lore of the Marsh. Pete is eighty-eight now, and doesn't get around so well as he did, but he remembers in sharp detail the whole process of haying, and he recalls the fires that came after the Marsh was drained.

We were in Pete's house at Marsh village, LeRoy; a real neat, white house and Pete, a large man, waiting eagerly for us, was lying on a couch. When we got inside and had sat down, Pete right away started to tell about baling hay on the Marsh.

I can remember back to when I was maybe five, six, seven years old. My dad had the first hay baler on the Marsh because the hay in those days in late last century was very thick and good. The grasses grew tall and sweet, and I remember well the smell of the hay as the mower cut into it. Dad bought first a horsepower baler, when the horses walked about in a circle and a tongue was fastened to the gears and turned then and ran the baler. The horses got very well educated and would go around by themselves. Nobody had to drive them or anything. Then in 1910 dad bought a gasoline-powered baler.

There was a fine market for the hay, because the breweries in Milwaukee, they was lookin' for that kind of hay. They used it to pack bottles into big cases. They twisted a bunch of hay around each bottle and they would set the bottles in upright, or stick them in top down. This is the way the big breweries in Milwaukee packed the bottles of beer for foreign shipment. It was the way, I guess, that Schlitz shipped that cargo of beer I heard they sent down to Cuba to greet Teddy Roosevelt's Rough Riders when they come tearin' down off of San Juan Hill! We shipped the hay into Milwaukee by the carload, and it was said to be the best packin' hay they could get. My dad baled most of the hay.

Sometimes we had to do some winter haying, but not very often, only when the breweries ran out of hay and needed it bad. If you bale hay in the winter time and a frost comes like we had on the Marsh the other day (1972) the frost will stay on the hay and when it is baled it will turn into a chunk of ice. Hay we made in the summertime, and then we stacked it. Stacks of eight, ten ton in a stack, then we would bale it later.

Peter Feucht, Pioneer Marsh Hay Baler

The Marsh hay was so good for packing beer because it had no "sticks" in it. It was all fine stuff mostly, about eighteen inches to two feet in length. There wasn't any goldenrod or black-eyed Susans in the hay. This made "sticks" and wasn't acceptable to the breweries.

Nowadays, of course, the breweries don't use Marsh hay any longer. They have all gone to packing with paper tops, and so the Marsh hay business is gone, now. In addition to the hay that Dad sold to the breweries, he sold wagon loads of hay to farmers or livery stables, for two dollars a load. They hauled a lot of the hay in the winter, and some of it went out west, too, to save the cattle from starvin' in bad winters. It wasn't much good as a milk producin' hay. Nothing like alfalfa, but would keep cattle from starvin'.

The breweries paid my dad five dollars and a half a ton. My dad always said he could make a little money by gettin' that price for it.

The kind of baler we had, well, you dragged it up to the stack of hay. Then we had to prepare the wires to tie around the bales, you know? And we had to make a loop in the end of each wire. It was quite a job just getting the wires ready. And the whole thing: the tall stack of hay, with the sides of the stack sort of slick, because when the hay gets pressed down in, it is like the thatch on the top of a house, you know. With all the stems of the hay coming down, understand, and that turns the rain. And there would be a small odor around the stack. I don't know. Sort of faint like the Marsh itself on a summer day . . . or like it was a summer day remembered, you know? And we would be there, all us boys, and dad, who was the boss, and the teams, ready to go on the baler, or if it was the gasoline engine baler, it had a big flywheel and a belt pulley on the side, and dad would always be fussing with the engine, and finally he would get a hold of the wheel and give it a pull or several times he would have to crank the flywheel and finally she would pop and some gas smoke would come out and then she would settle to a steady popping, because it was only a one-cylinder engine, you know. And us boys would get out pitchforks, and dad would climb down under the baler, because that was the way you had to do it. You got down underneath to tie the wires, all by hand.

Now I am eighty-eight, and I can see, and I can remember those smells on a day we were baling hay. The way the engine and the hay and the horses all mixed up and smelled together. And I can remember the way my clothes felt, too, the stiff overalls and my rough shirt sort of softly scratching on my arms, and my cap that had a bill on it and the smooth feel of the handle of my favorite pitchfork . . . because all us

boys had us a favorite fork that we used, and of course, the handles get smooth like silken wood, and dark a little with the sweat of a fellow's hands, you know.

In 1910 the demand for the Marsh hay was 'specially great, and that year there wasn't a place on the east side of the Rock River; you know how the Rock she comes down through the Marsh, in two branches, the west branch and the east . . . well, on the east side, not a spot of hay that wasn't cut that year. That's how big the demand for Marsh hay was. There was a hundred carloads of hay shipped out of Mayville in 1910, just to Milwaukee, and all that hundred carloads went through dad's baler. Oh, those were fine days, boys. Just the feel of walking through a Marsh

Haymaker On The Marsh

hay meadow, and as you walked, you would sometimes scare up birds and rabbits, and the mother birds, larks, would get up a little way crying, afraid that we would hurt their nests. Dad was always as careful as he could be about hurting the birds and the game. I don't know, we were all like a big family someway. The birds and the men and we sort of had a common feeling, I guess. It was like a sort of unworded love, if you wanted to think like that. Sure we hunted the birds when the time came, but we loved them, too. And we never hurt one while we were haying if we could help it.

And we had a bull rake, that was a big, wide rake with long spikes or prongs of wood sticking out for maybe ten feet or so . . . thirty of these long wooden spikes, maybe, and attached to wheels, and the team pushing from the back . . . we would rake the hay into windrows with a sulky rake, a smaller rake with a trip lever, and then the bull rake would come down the windrow and push right along the windrow until the hay was piled up and up, almost a mountain of hay it seemed to us, and the horses would push it up to the stack and there we were waiting to stack the hay . . . And if the day was hot as it was, usually, and

Boy With Coons

in the Marsh meadows it would seem especially hot, with the birds singin' and all, we would be sweatin' and it seemed like we were just overpowered by all the hay . . . no end to it.

Dad owned only about sixty acres of Marsh hay land himself, but he bought the hay from all the rest. He paid from about seventy-five cents an acre up to the two dollars an acre depending on the kind of grass.

In those days there were thousands of prairie chicken on the Marsh. But about the time I started to hunt . . . the prairie chicken all left.

Eddie Lechner, who had been itching to talk, said: I can remember the last flock of prairie chicken on the Horicon Marsh. And I can remember the last flock of quail. Seems impossible that they are all gone now. There were thousands and thousands. I remember the year because it was the year they built the golf course at Mayville, and it was built by PWA labor . . . hard-times workers. All us guys were lookin' for work. I was a young married man. Some of us were starvin' to death. I had to go with a truck out to George Feucht's farm, and we had to get this peat, to make compost for the greens. We did this in the winter time. And there was still a flock of thirty to thirty-five prairie chickens down there, over by the pines. We saw them and they came out and ran. How they could run! 'Course, those days it was all brush. Wasn't brushed out like today. Well, that winter we had a little snow storm, and we drove out that way toward George's place, and you know how the quail will huddle together in snow to keep warm . . . they will find a place under a bank or where the grass has held the snow up a little above a log, and they will gather under there and get together and keep warm; maybe a hundred will all be together there . . . We stopped, I remember, and sat there in the truck and watched the quail who were right there by the road, and we could see their heads moving. If we had gotten out of the truck I guess they would really have exploded out of there. They come out like a bullet, you know. That was the last quail I ever saw around the Marsh, and that was the last flock of prairie chicken. Old John Postion was with me that day, and old Mickey Haney. He was the dynamiter, and we also had to get sand from George's pit.

Well, all of that old stuff was changed when the dam was put in at Horicon, and the Marsh started flooding more, and the hay business went out. Prairie chicken like a hay meadow. Aren't hardly any now.

Pete Feucht said: Yeah, chicken like a cover about twenty inches high. They can hide in that. I was walk-ing on the Marsh one day in the old days, and I come on a chicken with young . . . they was so small that they could hardly run, and I thought I could maybe catch a few of them chicks and take them home and raise them, you know. But they was gone into the cover before I could hardly blink. I couldn't find a single chick.

The storms and fires were sometimes terrible on the Marsh.

In the spring of 1889, little rain moistened the earth and for more than two months not a drop fell. Upland meadows wouldn't hide a rabbit. Farmers, some from as far as twenty-five miles away, flocked to the Marsh to make hay. It was a hayrack parade on country roads. At six in the evening on Sunday, August 9, after a sizzler, a combined cyclone and twister struck the northern part of Dodge County. Hurricane winds twisted huge oaks like ropes and lopped them over. Many of the trees were uprooted and rolled away. For four hours after the winds subsided, the rains came in torrents; then the skies cleared, the world was at peace. On the following morning, in the short space of fifty rods north of the Chester Town Hall, twenty-six loads of hay were bottom side up — wheels in the air — sight seen on all roads for miles. Familiar oaks were missing. The entire roof was off Abel Wright's barn. Grain stacks had been torn down and bundles of grain scattered to the winds. The drought and the storm were harbingers of the end — the end of good, clear haymaking on the Marsh.

Late that fall the dry peat caught fire from burning grass. The top six inches of soil burned on large areas, and thereafter nothing but weeds and nettles grew. In other locations the whole forty-two inches of peat were consumed by fire. As a result, potholes formed, many irregularly shaped and some long and straight, with large bays on either side made as the wind direction shifted. In a few places the peat smouldered all winter beneath deep snow, and the air was filled with smoke and bad odors. After the winter snows and the spring rains had partially filled the holes with water, wild rice and other vegetation that wild ducks crave took root. In many places tall, thick grass grew to the water's edge, making natural blinds for hunters. These holes made the best of hunting grounds.

Pete said: I can remember, too, the big peat fire of about 1902. But that fire only burned on the blue joint bogs. When it come to the Marsh itself it didn't burn. But in 1933 it went over the entire Marsh, everything burned. All of the vegetation including the roots of the hay. Everything. It was a terrible dry year. Well, it was dry I guess all over the whole country that year. We were baling hay at one of the farms in the Marsh, and when the fire come we pulled the baler away from the stack, and the fire run kind of around the stack, because the farmers had burned around the stacks, you know, to save them. And the fire came and burned into the peat of the Marsh which was dry then, and it burned down into the peat three or four feet and more, some places.

The fire turned the whole Marsh into a non-productive land. But it did have some good. It burned down into the peat, and made potholes and these became, when the water and new growth came back on the Marsh, great places for the ducks to feed. Maybe that's one reason we get so many thousands of ducks back on the Marsh. When the big fire came, the peat holes from the past fires, you know, had just about grown even again. You could walk over them. That's the way the peat accumulates. The weeds grow and rot and drop down, and just keeps building up. Now, most of the hay that's cut is cut on the west side of the Marsh. There is a man in Beaver Dam named Carp who is in the hay business and he has a baling crew. But that hay that he cuts is used for blowing onto hillsides to hold the soil when they are building highways. So that's another use that the Horicon Marsh hay has nowadays. And I heard that they do use some of the Marsh hay now for packing fancy wine bottles. It seems to be especially good for that, too. But there isn't acre and acre of hay in the Marsh like there used to be. You have to cut it just in spots, now. But some farmers do still cut some hay in the winter time for bedding for cattle. They don't feed it anymore like they used to.

The big fire of 1933 improved the Marsh for the wildlife. Because the potholes filled with water when the State of Wisconsin finally built the dam at Horicon in 1930, and that made feeding places for the ducks; and when the big fire come, anyway, there wasn't much wildlife, because the whole Marsh was so dry, you know. There was still water in the ditches and that's where the geese and ducks were. Not on the Marsh itself that year.

My uncle and I we hauled hay to Horicon. We tied the wagon in back of the sleigh. We had two teams. He drove a team and we had about two and a half tons on the wagon. We hauled about five or six carloads. It was the balance of the hay that we baled that winter.

When the big fire of 1933 came, Lehner's barn over in the Marsh was in danger of burning. Lots of families in the Marsh had to move clear out. But we had a little fire department in LeRoy and we watched that barn for two weeks. Night you could see them sparks fly. Daytimes you couldn't see 'em fly. We watched that barn, and Lehner had a straw stack close to the barn. And that barn was pretty close to the Marsh. We knew if the straw stack would catch afire, since it was closer to the Marsh, the barn would go. So the fire insurance adjuster called us one night about eleven o'clock. The wind had shifted to the east, and he says we ought to do something with that straw. He was afraid the straw would fire and then the barn and his company would be stuck, you know. So he was watching close, too. Well, we just pulled a hay baler out there in the night and baled that straw. That's the way we got rid of that.

I used to trap some, too, and I used to get muskrats through their house. One Sunday our family caught over a hundred "runners." A "runner" is a rat that has run away from the muskrat house in the winter time when he can't get to his feeder. You see a muskrat, he has a big house . . . you've seen them, hundreds of them on the Marsh. And maybe about fifteen feet away he has another house, like a bushel basket, and this he builds up with food. And he will swim from his dwelling over to this feeder. He grabs a mouthful of food and brings it back to his house and eats it in there. But if the ice freezes down, he is going to begin to get hungry and starve, so he gnaws

From Pen To Freedom

out of the house, and when he gnaws out of the house he is a little too stupid to find his way back in, and then the rats are running all over the Marsh. And it was these runners that we got more than a hundred of, one Sunday. One cold winter, I believe it was the year 1936 there were runners all over the whole Marsh. It will happen any winter when we get a freeze down. They come out and get on the snow drifts and when the spring thaw comes there will be eight, ten inches of water come in the ditches and that water will be full of rats!

Ed Lechner said: One Sunday a family called me and asked if there was still an open season on rats. I told them, I didn't know just offhand, but would let them know. So when I got back up to Marsh headquarters next day, I mentioned this to Jim Bell, the director, and he says the season is closed. Well I knew something unusual had happened, and I went over to see these people who had called me, and they had a pile of runner rats about so high . . . maybe a hundred or so; well, they were out of order. But the runners would die anyway. And the family never sold the rats or nothing. But that's the way things happen sometimes.

Pete Feucht said: Back about 1910 we were getting eleven cents apiece for rat skins. But at that time you couldn't earn more than a dollar a day in wages. It was a good day's wages if you could take ten rats, well you had a day's work. We hauled the rats, or shipped them to Percy in Oshkosh. A big fur house. A rat today, 1972, will sell for about a dollar.

In 1938, the Marsh was flooded and that was one of the great duck years we ever had here on the Marsh. You could stand almost anyplace and shoot your limit of ducks. My grandfather owned the home place on the Marsh by inheritance. My great grandfather had it first. He lived out there. He came from the old country, Germany, of course, and my grandfather's wife died on the way over. He had four children with him and he come out from New York by horses. Some of them kids are buried out here.

One day my brother is going to Chester over the ice. And he is hauling wood. And his horse went through the ice. Well he run over to Strooks Hunting Lodge. It was a place where rich men come to hunt, you know, and there he found several men playing cards. He hollers, "Can you come and help us? Our horse dropped through the ice into a pothole!" The whole gang come running out to help; and in a few minutes we learned how to raise a horse that had fallen through the ice. An old Indian showed one of the men how to do this. He was an old Indian who used to live on the Marsh and was always hauling

wood. You have to choke a horse that has fallen through the ice; being careful not to choke him to death, of course. The wind will stay in the horse's lungs while you are chokin' him, of course, and cause him to float, and then you can slide him out onto the ice. This is the way the old Indian done it, and it was the way my brother and those men from Strooks Lodge got the horse out of that pothole!

We used to be able to pick up any number of Indian relics around the Marsh. Some stone, some copper. We had a big collection of these things until my sister told us we ought to give them to somebody who knew about them. We did. Never had a word from him again; guess he started his own museum!

Ed Lechner started then to talk about strange lights in the Marsh:

When I was living in Kekoskee, Victor Schwartzmiller worked for the Collien Boys. Vic was walking home one night and he looked around . . . he was on a Marsh road, in a low spot, and he looked around and here a ball of fire was following him.

He was scart to death. He run all the way back to

Kekoskee, and when he got there he run into the house where his folks lived and he was white. He told how this ball of fire had followed along behind him, and how it got bigger, and closer, and he run faster until he said his feet wasn't touching the ground at all. He run through swamp and potholes and everything. He never stopped for nothin'. He run through the brush and when he come to a tree it had better get out of the way, that was all, or he run right over it. He says how he could feel the heat from this fireball scorching his pants as he run, and he says he knew it was the devil back there just aplaying with him and tickling his tail and back with the heat. He claimed he could smell something like brimstone and he knowed that for sure then it was the devil.

Well, when he was telling about it a neighbor happened in and he seen Vic all white and shaking and his clothes all wet and ruined. He heard what it was about then he started to tell how it wasn't the devil at all but a bunch of this methane marsh gas that when it comes like that will glow in the dark like fire. Ghost fire he called it . . . some say will-o-the-wisp, or jack-o-lantern. And a lot of people from Chester has seen it.

We drove out, Uncle John and dad and I and the family. Uncle John had this team which he used for hauling beer and we went out looking too, for this ghost fire. We didn't see anything. But Vic he swore he saw it, and he wouldn't never believe it wasn't the devil. He swore it wasn't no marsh gas. Said he felt an evil somethin' there. Made a man run fast, he said.

Pete's wife, Mary, said:

Well, I been out in the woods a lot of times and I saw a light there that would go along in front, skipping and hopping along, long and short, back and forth, and it was all eerie and white and I saw that many times when I was walking in the woods. I heard tell that it was rotted wood that makes that . . . That was in Fish's Woods, and you could always see those lights walking in there. I ain't sure that it wasn't spooks, even yet. When I was nine I was scared to go in there.

Yes, says Pete, there was a feller from Chester one time, and we were fishing out there and we seen those lights . . . a lot of them. He was a good Methodist and he wouldn't have none of that marsh gas stuff. He claimed it was the real thing . . . whatever that was.

There was this feller named John Tiedeman who lived down here where Frankie Shaw lived, and he had a farm across the Marsh; and he was gonna drive his cattle across the Marsh. So he took a muskrat spear (it was a long, sharp iron rod) and he was going to test the ice. He was carrying the spear loose. I suppose when you get cold you don't want to have it held too tight, and he fell and run that spear into his guts, and he walked two miles on skates for help . . . there wasn't no telephone then and they had to drive a horse to get him, of course; they operated on him, but he died.

Pete says:

When there was a strong wind the fires in the Marsh would travel faster than a feller could run. Once I was up near LeRoy and a fire got started at the "Y" where the two branches of the Rock River come together down near Horicon. And it traveled about twenty miles in an hour. That's how fast the Marsh fires went. And then, of course, the peat would burn for days and days after that.

If you got caught out there in the Marsh with a fire

coming, you lit the Marsh ahead of you . . . a back-burn they called it, and that would maybe save you. It was the only protection you had.

My dad came running into my bedroom one night yellin': put your clothes on. We got to go down and save five stacks of hay. A fire is comin'. He says, get up quick and run down and light the fire in front of the stacks. When I got down to the first stack, I was choking from the smoke. All I could do was hold onto my nose and cough. And I set the back fires around the stacks and saved them all. The way I set the back fires, I would scatter a thin layer of hay all around the stack, as far back as I wanted to burn. Then I would set this afire. In 1910, during a fire, it was so light all through this country around the Marsh that you could read the paper, at midnight. And the whole thing was a spectacle . . . great shoots of fire coming up and a glow over the whole thing, and where the peat was burning you could see a kind of low glow coming from the earth, and the water, what there was, would glow, too. And you could hear the fire cracking, and you sat there and thought of the poor animals caught in it, and wanting to get away . . . But those days we didn't have any deer in the Marsh. Now we do have lots of deer, but them days we didn't.

Mary Feucht says:

We used to have parties, the neighbor kids and us, all the bigger ones, and then we had a fire once when we were having a party and they all run to the door, and it was so light we all thought it was just like morning, and we all wondered how we were going to get home, and the ashes were falling all over everything.

Pete says:

One of the Schultz girls taught me to dance. And not long ago I met her and she says, Hey, Pete, she says, I danced with you already. And I says, it must have been a long time ago, and she says, sixty years, when I taught you the polka!

Proud Mother And Her Family

THE DUCKS

During the spring of 1910, Wilton Erdman, a Horicon boy, was standing on the corner of Lake Street and Vine. He was ten years old and a very sharp lad. He was standing on the oak plank sidewalk, and he ran his toes between the cracks of the wide boards. The cracks were large enough so that he could get all five of the toes of his right foot into the crack; and as he stood, experimenting with the feel of the rough and cool boards, for it was an early evening in late April, the village lamplighter came to light up the gas lights with his carbide lighter fastened to the end of a long stick.

It had been a rainy day and the street was still wet though now the low sun had broken out and the sky was a wonderful light green with bright clouds in the west. The last of the beer wagons from the Fox Head Brewery at Waukesha that came almost every day, or the wagons from the Ziegler Brewery at Mayville and even the little Groskopt Brewery at Horicon, rumbled down the street and disappeared. The beer wagons with great draft horses straining at the traces and the barrels set in on the slope against the center boards, rolling a little, was a favorite sight to Horicon folk.

The people who passed along the walk were mostly going home for supper, and Wilton knew that in a little while he too would have to go home. But he was waiting to see a very interesting and thrilling sight. It was, he thought, time for the grand passage of the ducks to come, and a show he could not miss.

Then, suddenly, there it was . . . that strange and exciting sound from far away it seemed, and then quickly, closer and closer until it was like a kind of rolling thunder. The boy waited, gazing ecstatically into the sky, the marvelous cloud of ducks filled his whole vision. Flight after flight. Thousands and thousands! And the ducks went over the town and into the Marsh where, the lad knew, they would rest awhile before they migrated on.

The boy thought there must be millions of ducks. That was the same year the great dredge, owned by the MacWilliams Drainage Company, was doing work of digging the main ditch down the entire length of the Marsh; and the ditch was to begin the drainage that after a while would make the Marsh uninhabitable for such vast flights. The boy didn't know this on that evening in 1910; but he found out later, and when he joined Curley Radke, the Horicon lion-hearted crusader who fought, eventually, for the restoration of the Marsh. The boy Wilton Erdman never forgot the thrilling flight of the ducks on that September evening in 1910.

Ducks have always been of supreme importance on the Marsh. In the Indian days the ducks were said also to cloud the skies like the passenger pigeons. The white settlers found thousands of ducks when they came, too, and in the great hunting club days, sport was chiefly a matter of shooting ducks. There literally were no geese at all. When the drainage project took place and the Marsh, for a time, became a sort of desolate wasteland in the period from about 1914 to 1930, the ducks almost disappeared.

When the Izaak Walton League of America set out to get the swamp reclaimed as wetlands, and were successful, finally, Curley Radke, the spearheader and fearless worker of the "League for the Marsh" project, believed that the ducks could be brought back.

I talked one afternoon with this same Wilton Erdman, now a Horicon old-timer, about the great duck renewal project on the Marsh, now folklore . . . and of course we covered a lot of other items as well. Wilton is kind of laid up now . . . he lost a leg through an operation awhile back, and he has to get around his large house in Horicon in a wheelchair. But he's exceptionally keen, and his house is a treasure mine of Indian relics, and boxes filled with notes and clippings which Wilton hopes to assemble into books some day.

He speaks quickly, eagerly, as though he can't wait to get from subject to subject. As a lead-in to this whole duck story I asked Wilton what the purpose of the famous and now, somewhat legendary, "Duck Liberation Day" was . . . when it was held back in 1934.

Wilton told me that Curley Radke organized the "Duck Day" to get the ducks coming again. They didn't have many ducks comin' back in the early 1930s, said Wilton, and the ducks were not being propagated. Radke got all kinds of ducks together and let 'em go . . . all at the same time . . . domestic ducks, wild ones . . . everything. And a lot of those domestic ducks wanted to walk through the town, and did, and the hunters shot a lot of 'em, and a lot didn't survive the winter. Those that were wild, though, did stay around and they actually got a few ducks started. Radke was the originator and the hero.

Hoping to find out more of the Radke story I asked Ed Lechner about him and was shown on the grounds outside the office building a large stone marker honoring Louis, or "Curley" Radke. The marker said:

> Radke Memorial, dedicated to the memory of Louis (Curley) Radke 1884-1947. For a lifetime devoted to Wisconsin Conservation and restoration of the Horicon Marsh. A true defender of soils, woods, waters and wildlife by Wisconsin Division of the Izaak Walton League of America, erected 1950.

Radke, I found, more than any other one individual was responsible for saving the Horicon Marsh and for paving the way for the creation of the Refuge as it is today. He led a successful fight to get the Marsh flooded again through the rebuilding of a dam at Horicon. In appearance he was a large man, deter-mined, and fearless.

He mustered the entire organization in the Izaak Walton League behind his idea, and the Horicon Marsh Game Protective Association was started for the purpose of carrying on the fight. Opposing, of course, were the drainage interests and owners of the Marsh lands, and a small number of farmers who saw their own lands threatened by a flooding project.

In less than twelve months the Horicon Marsh Project became one of the leading causes of conservation-ists in the United States. Most of the service clubs, chambers of commerce, Federated Women's Clubs, Rod and Gun clubs all endorsed Radke's reflooding plan.

On May 24, 1924, the President of the United States, Calvin Coolidge, held a national conference in Washington on recreation and outdoor life. At the conference the Horicon Marsh Project was presented as a foremost cause. American wetlands must, the conference voted, be preserved.

Radke circulated petitions. More than 100,000 citizens in Wisconsin signed. The idea was to gain as complete public endorsement as possible. They fought the issue through seven consecutive sessions of

Fond du Lac Hunting Club Of Yesteryear

the Wisconsin Legislature between 1923 to 1930. The battle to flood the Marsh was carried to the Supreme Court. In 1927 the Waltonians secured the passage of a bill placing the Horicon Marsh under the jurisdiction of the State Conservation Commission, to be restored to a wildlife refuge for all the people. The law said, in brief:

(1) A wild life refuge, game preserve and fur farm shall be established on the Horicon Marsh in Dodge County under the supervision of the Conservation Commission.

(2) The Conservation Commission shall purchase or acquire by condemnation proceedings the land known as the Horicon Marsh, or as much thereof as it deems necessary, and may construct such buildings thereon and provide such equipment as is reasonably required to carry out the purposes of this section.

(3) The Conservation Commission is authorized to construct and maintain a dam or dams in or near the city of Horicon to control and regulate the flood waters on Rock River, and to restore the public waters of Rock River on Horicon Marsh to the natural levels existing prior to the private drainage of the same.

At the same session Chapter 475 and Chapter 479 of the Laws of 1927 were enacted, the first appropriating the sum of $25,000 ($25,000 annually for ten years) for the establishment of a wild life refuge on the Marsh, while the other appropriated the sum of $10,000 for the building of the dam.

Just how impassioned was Radke's battle is shown in an article he wrote in 1925 for *Outdoor America.* The old Marsh, wrote Radke, is calling to its hosts of friends — calls to them for release from suffocating weeds and Canada thistles, from a vampire stream (the main drainage ditch), a drainage scheme that has destroyed the beauty nature once intended for all.

If the activities of a few drainage fanatics, who care nothing for public rights go unchallenged, then the old lake bed's virgin loveliness and fragrance, its golden sunsets and shimmering moonlit nights, its thrill of wild life, and its reedy, wave-washed banks, will be lost to the enjoyment of thousands yet unborn. Waltonians ask only that man be compelled to restore the water he has so ruthlessly drained away.

There is an awakening call in the breezes, there is a challenge in its possibilities. Horicon Lake is not a dream of the past, it is a real hope of the future. It is the out-door heritage of posterity. Here is a plea, coming quite unsolicited and entirely without suggestion, that should go straight to the heart of every legislator, every adult of these United States, every mother and father.

Read it, think about its naive, simple, almost terri-

fying truth, in face of reckless American destruction of the "heritage of the children."

And the children of Brandon, Wisconsin wrote Curley Radke a letter:

We were interested in articles about the Horicon Marsh because we are near enough to it to be vitally affected by whatever action is taken in the matter of flooding these forty thousand or more acres. We, as children of Wisconsin are interested because we are the ones who will receive the most important benefits, should the Marsh be made into a lake once more.

The heritage of our grandfathers was the herds of buffalo and the flocks of pigeons. The heritage of our fathers is the flocks of geese and ducks, and the muskrat and mink. The buffalo and passenger pigeon are almost extinct. If the animals of today are not protected and provided for, what shall be left for us?

Louis "Curley" Radke

Radke Memorial Dedication

We, as representatives of all the children of Wisconsin, ask to have the Horicon Lake restored. Voters of Wisconsin, you owe it to us, the children of tomorrow!

(Signed) The Kolstan School,
Brandon, Wis.

Curley Radke continues:

If the grade children of Wisconsin realize the possibilities for an out-door future in the restoration of these lonely acres, what can grown man do to make this realization a truth? Fight! Fight for the old heritage, the sight our fathers saw when, with rifle in hand and sturdy hearts, they gazed over the waste land and saw in their mind's eye a lake with waters blue, and wild fowl nesting at its weedy brink.

"Does not the voice of reason cry
Claim the first right which nature gives,
From the red scourge of desolation fly,

And claim our heritage for we who live."

There is no more to say. We fight not for the dollar, not for a name in the halls of fame, nor for the glory of man or state, but for the millions to come, the tomorrow of our boys and girls.

Now Radke proposed a Duck Day to highlight a return of ducks to the Marsh. To make the day successful, Radke enlisted help from all over the United States. Among those who eagerly responded to his "duck renewal" was Joe Penner, radio and motion picture comedian whose famous line, "Wanna buy a duck?" was in those days being imitated by nearly every other comedian and by lots of average citizens as well.

To restore wildfowl to the great Marsh became Radke's greatest hope. He argued that migratory birds have traditional stopping places along their migration routes or flyways, and that the knowledge of such routes is passed along from one generation of birds to another, since the young birds travel in company with the old. If the pattern of stopping at a particular place is interrupted, then the birds forget about it, and they no longer stop. Finally in 1930, when a dam was again built at Horicon, the Marsh started its long comeback.

Virginia Palmer, writing in the *Wisconsin Magazine of History* for 1962-63, Vol. 46, relates: Louis Radke conceived of a (bold) plan to bring back the enormous numbers of birds to the Marsh, just as there had been in former years. The Conservation Commission agreed to lend its support in furthering the plan — "planting" domestic ducks on the Marsh to attract wild birds on their spring migration and to establish the Marsh in the minds of the birds as a desirable place to rest, thus insuring their return in the fall on their return flight. Radke explained his plan to civic and educational groups throughout the state, crusading tirelessly for the future of Horicon Marsh. It was plain that the success of the plan would depend on the interest and co-operation of a great many citizens. Joe Penner generously agreed on his regular Sunday evening broadcast to publicize the need for contributions of pairs of ducks. As the idea caught on, so many people did want to buy a duck that it was not long before almost every farmer's barnyard in Wisconsin was emptied of ducks.

Well in advance of "D" Day the Conservation Commission released several hundred muskrats on the Marsh to do their part in making it more attractive to the birds. In building their homes underwater, muskrats throw up mounds of dirt which rise above the water to expose green shoots of water plants. These little islands, called "eat-outs," create ideal resting and feeding areas for ducks. The forestry service donated two tons of wild rice for duck food.

Duck Liberation Day, as April 20, 1935, was called, put the little city of Horicon, as well as Horicon Marsh, on the map. A crowd of 5,000 came from all over the state to bring ducks and see the fun. An active interest in the proceedings was taken by J. N. (Ding) Darling, head of the United States Biological Survey in Washington, D.C. Darling, a talented cartoonist, also designed the first federal duck stamp.

General Ralph Immel, head of the State Conservation Commission, attended the ceremonies held during the morning and expressed his approval of the experiment to the crowd. Conservation wardens from throughout the state were there to watch the success of what they realized was an innovation in the field of conservation. While the crowd listened to the speeches in the park, men worked feverishly to band the ducks that had already been contributed in order

to trace the success of the experiment. A total of 1,180 domestic ducks and two wild Canada geese were banded that morning. Twenty-five high school boys drove up from Milwaukee at the last minute, each bringing a duck. One duck was shipped all the way from California and another from New Mexico. Ducks continued to come in all during the afternoon and many people said that they would have liked to contribute a duck but that there were none available for sale in the state. Even Joe Penner, who had promised 500 ducks in addition to his effective promotion of Duck Liberation Day, had only been able to obtain 200 and had to be content with sending a check for the rest.

When at last everything was ready, the band led the boisterous parade to Quick's Point jutting out into

heaved a sigh of relief as the ducks swam contentedly about. Duck Liberation Day was a success.

Meanwhile, at Mayville, on the eastern edge of the Marsh, a disgruntled group of landowners, the Horicon Marsh Farm Land Protective Association, was holding its own demonstration. The speeches, unlike those heard at Horicon, expressed great dissatisfaction with the state of Wisconsin for not compensating the owners of land for damage done to crops by the reflooding of the Marsh. Wooden duck decoys were floated on the Marsh in mock imitation of the duck liberation over at Horicon. But few, if any, of the jubilant conservationists paid any attention.

A few months later the 236 land owners took more drastic action, petitioning the circuit court in a suit against the State for an order to open the dam gates and lower the water level on the Marsh. The judge ruled in favor of the land owners and the dam gates were opened, draining away the water at a rate of one-half inch per day. The water was eventually to be lowered a total of four feet. Soon (once again) thousands of pickeral and other fish lay dead of the Marsh floor, allowing the dreaded botulism bacteria to develop. Most of the living plants and wildlife on the Marsh were caught in death and decay. Old-timers who remembered that William Larrabee, founder of

the Marsh. There were shouting school children, noisy, quacking ducks in their crates, and Sousa marches played by the high school band. At the Point the first duck was released by "Curley" Radke in honor of Joe Penner, who had done so much to make the occasion a success. The next several ducks were released in memory of persons vitally interested in the venture but who had not lived to see the great day.

At last it was the turn of the impatient school children. Each child ran to one of the waiting crates and opened its door. The crowd held its breath and waited to see what the ducks would do when they discovered that they were free. After the first few ducks had tried to Marsh and discovered what an ideal place it was, the rest followed. The crowd the city of Horicon, had adopted the Indian name of (Lake George) in his native New York State meaning, "clear, pure water," shook their heads sadly. The Marsh was in danger of becoming a menace to the public as well as an unpleasant sight. Farms further down along the Rock River were being flooded because the Horicon dam no longer held back the water. At last, conservation interests petitioned the Supreme Court for a temporary order to close the dam gates and flood the Marsh again, at least until a hearing could be had on the matter.

In order to become better informed on the subject of the controversy, Governor Julius P. Heil, Attorney General John E. Martin, and several legislators made a personal visit to the Marsh in May, 1939, which convinced them of the importance of the restoration of the Marsh to its former condition. Finally the state assembled enough money to purchase 800 acres of land. Radke and Owen Gromme of the Milwaukee Public Museum, as well as other interested men and women, urged that the Federal Government purchase and control a portion of the Marsh area. The Pittman-Robertson Wildlife Restoration Program, an act passed by Congress in 1938, provided that an excise tax of eleven cents on every dollar spent for sportsmen's guns and ammunition should go to the State for the purchase of land, wildlife preservation, and management. Funds from this source were used to obtain a portion of the Marsh, while fifty cents of every duck stamp purchased by sportsmen for the privilege of waterfowl hunting was spent by the Federal Government for the same purpose.

Frank "Butch" Burkhardt, Old-Time Marsh Fur Buyer

126

THE RAT BUYERS

The old Indian peoples came to the Marsh partly for the fur. The muskrats were their brothers. The Indians noted the sturdy houses the rats built. And they prized the bright coats. There were over 100,000 rats a year that were trapped off the Marsh by white men, beginning in the 1880s and 1890s . . . how many were taken by the Indians? Yet the rats thrived and apparently could not be trapped fast enough to noticeably decrease their population. Frank Bossman, who lived in Horicon, bought a hundred thousand rats one year, and in Kekoskee lived Frank Burkhardt, another fur buyer. At the turn of the century, rat skins were selling for from eight to fifteen cents per hide. Later they went even lower. But to poor men fifteen cents a skin was good money in 1900. The only muskrats in the whole United States that are of better quality than the Horicon Marsh rats are those from Louisiana. They eat vegetation there that makes the fur even more thick and very beautiful; but Horicon rats are second!

Nineteen-forty-three was a good year. There was a lot of water on the Marsh and the State was buying up land from farmers for the Refuge. Barney Wanie went to the farmers and said to them: Why don't you let us close the dam this year. The farmers said OK, after the first of July, when the canary grass is all cut, you can close the dam.

Their willingness to have the State close the dam represented a big turning point. It meant that at last the farmers saw some good in having more water on the Marsh. And World War II was on, of course, and muskrat skins were selling very high. So the farmers

Sharecroppers Dividing The Catch

all decided that they were a bunch of royal fools for spending their time cutting hay when they could all be trapping muskrats. So the water went onto the Marsh and that fall, 1943, there were 120,000 muskrats caught, and the trappers, or farmers turned trapper, got $3.60 a skin . . . pretty close to a half million dollars in skins alone came out of the Marsh that year. The farmers harvested a lot better on rat skins than they made on crops that year, anyhow.

I had a pleasant visit in March 1972 with Frank (Butch) Burkhardt, son of the original "Butch," old time fur buyer; Frank Jr. is a retired tavern keeper and himself an old time fur buyer at Kekoskee. He is quite a guy. I went first to see him one very cold, snowy day in January and he wasn't at home. I parked in front of his comfortable home in the village, and got my car in too close to the edge of the road, and when I came back I was stuck, and royally at that. I had no shovel and finding one standing against Frank's front stoop, borrowed it and tried to dig out . . . unsuccessfully, but I did break the handle out of his shovel. I left a note pinned to his door, and went away, after being hauled out of the ditch by a State truck, feeling very guilty about Frank's shovel. I sent Frank back a check immediately for a new one, and in a few days the check was sent back to me with a note: Howdy Pardner. You don't owe me a damn thing. Breaking that shovel will keep me from breaking my back. Come and see me soon. Frank.

Well, that's the way things are up around the Marsh. I did go back and visited with Frank about a lot of old time events. He was telling me, for instance, that right after the first World War he was paying upwards of four dollars a skin for muskrats. They were in big demand, and a lot of fellows were trapping. The Horicon muskrats were also of exceedingly high quality which made them even more valuable. One month in 1910 Frank's father, who was almost the original fur buyer on the Marsh, bought so many muskrat skins that a whole large hayrack couldn't hold them all. He hauled enough muskrats down to the depot to fill an express car. A lot of the skins went to Berlin, to the Trousdale Company which was a chief buyer. Later they shipped a lot of furs to Oshkosh to Percy's.

Frank had a big warehouse for the furs which was pretty nearly always full in the season, and when they shipped the furs they sacked them up in burlap bags. The expressmen on the trains didn't particularly care to see the Burkhardts coming with all the skins . . . they would usually yell, "Well, here comes the Burkhardts with another load of them stinkin' muskrats!" Later, of course, Frank and his father used a Ford truck to haul the furs. But in those days anybody

could trap on the Marsh. "They didn't have to have a license or nothing," as Frank said.

Frank's father first started buying furs in the late 1880s.

It was a lively time when the big fur buying days were going on, and the trappers and many farmers who were doing some trapping part time would come up to Kekoskee with their rigs filled with furs . . . and sometimes they would get to talking and playing jokes on one another and sometimes the farmers would come with whole wagon loads of fish they had caught in the Rock River . . . which was literally swarming with fish of every species. Sometimes a group of farmers would suddenly wrap somebody up in a large coat or blanket, and while they were pretending to help him light up his pipe against a strong wind, other farmers would be stealing the fish out of his wagon. It was all part of a big joke: big country fun.

Sometimes they would remove the fish from a farmer's wagon while he was in the local tavern. Then when he came out and saw that his fish were gone, he would start crying and moaning that outlaws had stolen his fish. The tricksters would then tell him that if he would buy a few rounds of drinks they thought the fish would swim back. He always did, and they did.

Frank was in the fur-buying business and the tavern business so long he can't really recall when he did start in. He remembers that you could get six big beers for a quarter, and a free lunch, just as Ed Lechner said they had at his place. Only Frank said in the early days his mother would bake a large carp and set it out for a free lunch. You had to know how to cook the carp just the right way, and you had to know how much fat the carp contained, too, because the fat in the carp sort of basted the fish itself. Anyway, the German farm boys sure did go big for the carp, and now that carp are considered locally a no-good eating fish, it is hard to remember the days when the carp were really cherished by the local German descent Marsh people.

About nine or ten o'clock Frank's mother would bring out the big roasting pan with a couple of large carp all nicely roasted with a fine dressing in them, and the last sound that was heard was the scraping of the bottom of the roasting pan with the spoons by the hungry farmers who devoured every last scrap.

Frank heard old man Janke tell, a long, long time ago, that from the dam near Kekoskee when the suckers began to run in the spring, the men would gather on both sides of the river with dip nets on long poles, and all night long they would dip the suckers out of the river. They would smoke the suckers, and Frank says that smoked sucker is really very good. Also the farmers fed their hogs with the suckers, and there was a time around Kekoskee when you couldn't buy a piece of pork that wasn't flavored with fish! A story that is repeated in another way later on by the great yarner Emerson Hough.

A Horicon Marsh Fur Trapper Making His Rounds On Lehner's Ditch.

THE TRAPPER

Eddie Lehner, (don't confuse him with my guide, Eddie Lechner), old-timer, trapper and hunter, and occasionally, farmer, lives in a fine trailer home above the Marsh. His son, Gordon, does the farming now, and Eddie keeps pretty busy, too, but he doesn't trap any more. He was probably the most successful of the Marsh trappers.

Eddie was born right on the Horicon Marsh in a large, frame, white house that you can see for a long way if you are up on the hills at the east side. Bob Personius, the resident manager of the Horicon National Wildlife Refuge, lives now in the old Lehner place.

Ed Lechner and I visited Eddie Lehner. He is short, sturdy, nearly eighty, bright eyed, and can hardly contain the vast experiences he has had on the Marsh! Eddie and his wife lived happily in the trailer house after their son took over the farm. Unfortunately she died the day after Eddie's birthday in January.

Trapping, said Eddie, I done ever since I was six years old. I come from a race of real hunters. My daddy was one of the best shotgun marksmen that ever lived on the Marsh. When he was ninety years old he had never shot clay birds. But some fellers from Mayville, Bodden and them fellers, they wanted him to shoot clay birds, too. He had a double-barreled, so he went to work, and he shot with them. Out of twenty-five shells he broke twenty-two of the clay birds. And him ninety!

The most rats that Eddie ever caught in one regular season was about 3,000. But when he trapped special for the Government from the first of November in 1948, he trapped that year over a month be-

Eddie Lehner, Famous Horicon Marsh Trapper

fore freeze up, and he had 3,400 rats that one month. Some years he got over two dollars a rat. And some years he got over three dollars. The market would have paid more, but there was a ceiling on the price. There wasn't so many mink around at that time. But if you had a mink, even a small female mink, they would give you $100 or $150, to make up for what the rats were worth . . . even if the mink was worth only $10. Everybody was after the mink that year.

When Eddie was young he got six cents apiece for rats. Sometimes they would take as many as thirty-five rats out of a single burrow . . . a burrow will run in there for thirty feet maybe. You could sometimes scare them out and the rats would run out one after the other.

When the Marsh would burn, as it did sometimes, the peat would catch fire and burn away down from the surface; sometimes as much as twelve or fourteen feet, potholes, or burn holes, and the rats would sometimes gather in those burn holes. The biggest fire they ever had on the Marsh started the sixth of September, 1933. That was just before the State finally closed up the new dam they built at Horicon. Below the Horicon town line the Marsh had burnt the peat before, but there it didn't burn so deep; but at the northern end the Marsh started all at once that September, and the fire went right across those ditches. In three weeks the whole Marsh was burned out. The only way you can stop a peat fire is if you can smother it, and the ashes gathered on the surface of the Marsh, more and more, and finally the ashes got a couple of feet thick in some places, and the fire was smothered by the depth of the ashes. Generally, the great fire burned the Marsh from about eighteen inches to five feet deep. More in some spots, and then the willows all around the Marsh burned.

Eddie said that one year he and his brother trapped on the west side of the old Main Ditch. That was before the Federal Government came in about 1941 and established the Federal Refuge. At that time there was no limit on the number of rat traps that could be used. Eddie had about 300 traps out; his brother had probably 200. One noon they had taken several hundred rats and they were skinning them. Eddie could in those days skin about sixty rats an hour. Well, Eddie said that when they finished skinning the rats they would take their guns . . . he said there were some deep holes along the east side of the ditch where the ducks liked to gather. One of the holes was about fifty feet long maybe, and the mallards liked to light there and feed. Eddie suggested that when they were through skinning, they would go over to this hole and shoot a couple of ducks. They could hear the ducks already, when they left the place they were skinning the rats. Eddie said to his

brother that they dassant go any closer . . . the problem was that when the ducks were so close you would shoot them all, or maybe you wouldn't get a one. Eddie stayed a good gunshot away from the hole and his brother crawled up close; and Eddie's brother reared up and hollered "Whooroough!" and they went up, about two or maybe three hundred ducks were there all flying up at the same time. That time a hunter could shoot five shells and Eddie's old automatic blasted until it was empty. Eddie picked up seventy ducks.

Eddie wasn't married at that time, and when they took the ducks home his mother was dismayed. His mother says, what you gonna do with them ducks? She saw herself picking ducks for a couple of days, and ducks were no real eating treat in those days. Everybody had ducks, could have them every meal if they wanted to. But you get tired of roasted or fried ducks, too, no matter how fat they are. Well, said Eddie, we gave ducks away, whoever wanted some ducks, they just came and took them away. It was one of the biggest duck feeds our neighborhood ever had. Only Eddie got so damn much scolding from his folks too. They said: What the hell do you want with them ducks out there!

When Eddie was nine years old he bought a double-barreled shotgun. His dad gave him twenty-five brass shells. At that time there was no smokeless powder or anything like that; just black powder and Eddie loaded the shells, and by golly, when he had twenty-five loaded, then he went hunting. Twice a day. At first Eddie loaded the shells kind of light. Then as he got used to it he loaded heavier and heavier. The old black powder made a very loud bang, and the load was so heavy it almost ripped the gun out of his hands. With that kind of shell and that black powder he could only shoot one shell. Then he had to wait until the smoke had cleared away before he could see to shoot again. Sometimes when Eddie had an especially heavy load a streak of fire at least twenty feet long would come out of the barrel of the gun. That was because the powder burnt so slowly.

Early in the morning, because hunting always started in the Marsh about an hour before sunrise, wherever there was a hunter, you could see the big streaks of fire fly whenever a shot was fired. It was a real Marsh fireworks show.

When trapping season came, then Eddie made trapping his sole employment. That was the way he made money. As he said there sure wasn't any money to be made in shooting ducks. Once they got some commercial interests in the Marsh who were doing commercial duck and goose hunting. That was a terrible thing, Eddie thinks. But now things have changed a lot. These commercial hunters aren't here anymore, and the wildlife is protected. Good.

And once in a while an otter will be taken in the Marsh, by mistake. Ollie Shard caught one, and let it go of course. There are quite a few otter in the Marsh still. They aren't often seen, though. But one time Umpty, who trapped right down by the Marsh headquarters, he got an otter in a trap by mistake, and he went up to the headquarters and wanted somebody to come and help him. He couldn't get the otter out of the trap alone. An otter will fight like hell. So nobody from the Marsh headquarters would go down with him. And Umpty said: I can't get him out alone. There wasn't any other way, so he killed the otter. Then he took him over to headquarters. I guess they were sorry they didn't go to help.

Eddie never caught an otter himself, but about four years ago when he was out trapping, he had pushed his boat out about a mile from where he lived, and all at once he saw a movement in the water and saw a strange thing swimming . . . he knew well it wasn't any muskrat. It was black and big. It wasn't a goose, though there were plenty of geese around. But when Eddie had pushed up to about a hundred feet of whatever it was, he saw then that there were four of them. He sat and watched while four otters got out of the water and crawled up the bank of the ditch. He watched while they got back to the water and swam down about a hundred feet, and they raised up out of the water about a foot high and looked at Eddie. Then they began making noises and swam right up to Eddie's boat. They played around his boat for a while and finally they got tired, he guessed, of whatever game they had been playing, and headed off to the south.

The eleventh of November came the great Armistice Day storm as they have always called it. Year was 1940. It started out as a beautiful day. But the wind came up. Eddie's boys stayed on shore while he went out on the Marsh, and they shot and shot. He said the wind was so strong they shot two boxes of shells and never got a duck. Eddie shot seven or eight ducks that morning, and they fell right in a pothole that had been made from the Marsh fires burning down into the peat. He got the ducks out of the hole, but by that time the wind had come up very strong, and Eddie knew he better go home. He had better than a mile to go. The wind was behind him, and it blew the boat along so fast, Eddie said, that the prow was sticking up out of the water about two feet. When the boat hit the shore the wind blew it up onto the bank about twenty feet. He swears that this is actually so.

It was the worst storm and blizzard they had ever had on the Marsh. Some hunters froze to death in the storm. Like on Lake Winnebago, some hunters were out on the lake that morning. And they never made it home. And even on the Mississippi River it was very bad.

Eddie said that about twelve years ago, on the twenty-fourth of October, he got a stroke. He was going to go hunting that morning, but it rained a little bit and Eddie's wife said, Now don't go huntin'. It's rainin'.

Eddie said he guessed he wouldn't go. He could go hunting tomorrow. If he went out in the cornfield he'd surely get soaking wet. Eddie went back to the house. His son and the son's wife were there; Eddie was standing in the toilet when something hit him. He fell down of course, and was unconscious. They went and got the doctor and the priest, and the priest said to the family that he didn't think Eddie was going to make it.

They took him to the hospital at Waupun and he was there for ten days before he regained consciousness.

But the eleventh day he knew everybody. The first thing he knew, Eddie said, when he came to, he asked for his pipe. His wife said, Your pipe ain't here. Eddie said yes it was. It was in his overalls. Well, said his wife, your overalls ain't here. You know where you are? she said. You're in Waupun in the hospital. So Eddie looked up and he saw that the ceiling of the room was low. A lot lower than the ceiling in the old farmhouse at home. And he saw right away that he was in a strange bed and everything different than at home. Eddie thought then that he had to get up and he started to get out of bed but his wife said, No, you dassant get up. The doctor says you ain't to be up for a long time. Then he saw that his son Harold and his wife Gloria were there beside the bed. He got deter-

mined more than ever to get up because he thought he was going out on the Marsh to hunt and it wasn't raining, so he had to go.

Two days later, though, Eddie was feeling so well that the doctor did let him get out of bed.

After Eddie had been in the Waupun hospital for thirty days, he asked Dr. Ries when he could go home. Tomorrow? And the doctor said he guessed Eddie could go home. Eddie said to him that he had to get up to go deer hunting tomorrow. But the next day when Eddie was awake, early, because he had to get home to get his stuff ready for hunting, Dr. Friedrich came into the room and Eddie said, Where's Doc Ries? I'm supposed to go home.

Oh no, Dr. Friedrich said, Dr. Ries said you weren't to get out today. And Dr. Ries has gone deer hunting.

That son of a gun, Eddie said, so he goes deer huntin' and lets me lay here in the hospital. And Eddie felt real hurt that Dr. Ries would do to him like that.

The next day when the doctor came in, he was sad because he hadn't seen any deer, and Eddie didn't say anything because he knew that if he could have taken the doctor out they would have gotten deer.

Anyhow, it was the thought of the Marsh and all it 134

meant to him, Eddie says, that kept him going, and gave him the strength to pull through that stroke. He just couldn't wait to get back to the Marsh, to his guns, and his traps and his dog.

Back in 1948 that was a great year for trapping in the Marsh. The first trapping day that year Eddie caught 113 rats. His sons were trapping too, and the end of the first day they had totaled 487 muskrats. They started to skin the rats at seven o'clock and by eleven o'clock they were all done skinning.

But nowadays, Eddie will sit down and skin maybe twenty or thirty muskrats, if somebody needs help, and then he will get up and walk around for a spell.

The highest money Eddie got for a mink skin was $38. One year, old Butch Burkhardt came out and bought only mink. Eddie had twenty-eight mink on hand when he came, and they were about half big and half small. The small ones Burkhardt paid $25 for, and the big ones $38. He always carried cash with him. So Burkhardt took out his purse and dumped what was in it on the table. All twenty-dollar gold pieces. That was the most gold that Eddie ever saw. That was in 1923.

Eddie still likes to go out on the Marsh. In fact it's the biggest thing in his life. Once, a while back, when Eddie's wife was still living she almost fixed it so Eddie had to stop trapping. She met Bob Personius, Director of the Federal section of the Marsh at church, and she said to him, Mr. Personius, can't you fix it so Eddie can't trap no more?

Well, Bob looked around and found Eddie's son Gordon. He asked him if Eddie ought to be stopped from trapping, because of his health, of course, but Gordon said, why no. He's all right. Sound of wind and limb, Gordon said. And by golly, after that Eddie was the champion trapper three straight years on the Marsh. He beat everybody, young fellows and other ones too, though Harold Wagner was a mighty close second. Eddie knew that Harold was pretty close, so he didn't tell Harold how many rats he had caught. Well, Harold thought sure he had beaten Eddie, and when they got down to counting they found that Eddie had won . . . he had a few more rats than Harold. By golly, you beat me again? Harold cried. Now have I got to be beat every year by an eighty-year-old man! he yells.

Eddie says he could live from morning to night out on the Marsh. He had made up his mind once, not long ago, that he wasn't going to trap anymore. But when the time came, he just had to go and get some trap sticks cut. There is something about the Marsh that gets in the blood, Eddie says. It is a combination of solitude, just being there in the Marsh vegetation alone, seeing the different colors, being real quiet and hearing the birds and seeing the movement of the tall grass and the cattails. Watching the water move here and there as some insect or small animal or fish disturbs the water; and more than anything, studying about things as they used to be, folks you liked and knew well, and hearing them speak in the memory . . .

Memories Of Grander Days

hearing the sound of guns on a still morning in duck season, and sound of the wings coming closer, the beat of the wings and the sound of the ducks coming down. The low whimper of your dog, and the feel of your hands on the cold gun . . . all of that, and a lot more. Because a man who has been raised in the Marsh is a Marsh man, really, and can never be anything else. The whole feel and meaning of it, both past and the way it is now, lies there in his blood and his memory. He smells the food his mother used to make, cooking breakfast at the big iron stove, and the way the rain pounded on the sides of the old house, and the way the snows piled up and shut the roads and hemmed the family in, living on what they had stored up, reading whatever they had. Talking. Just liking being at home.

That was the way that Eddie put it.

For geese hunting Eddie isn't so keen. It is just too easy to get a goose around the edges of the Marsh. Last year Eddie didn't even get a permit. It's just no fun anymore. They don't give the farmers any more right to hunt geese, incidently, than they do the gen-

eral public. And this makes a lot of the surrounding farmers mad because the geese do a lot of damage to their crops. Yet the farmer can't hunt them . . . well, Eddie isn't so mad about the permit scheme. But he just doesn't enjoy goose hunting around the Marsh. There is no skill required at all. There are usually so many geese that a shot fired almost anywhere will bring one down. And the truth of the matter is, you can hardly give a wild goose away around the Marsh. Eddie's son, Gordon, doesn't hunt geese at all either. Gordon's son, about eleven, is really now the keen one on hunting. Eddie bought the boy a little .410-gauge and a box of shells, and at first the boy wanted to shoot at everything that was around. Eddie had to teach him about hunting and all that goes with it.

Eddie went out with the boy one day, and Eddie didn't even take his own gun along. But they were walking along and a goose flew over and the boy knocked it down. The bird came down in the Marsh and they went to find it. Eddie thought it was terribly important for the boy to get the bird, of course, and they couldn't find it. Eddie didn't want to go into the tall weeds himself because he hadn't been feeling so well. But the boy plunged right in and found the bird. Eddie thought that was a good first lesson. He has had the .410 drilled out so it will take a three-and-one-half-inch shell. It is a real nice gun, Eddie says, but isn't much of a goose gun. He may get the boy a .20-gauge soon. The kid is really gone on hunting, as most of the young fellows around the Marsh are.

Eddie is retired now, and it was a couple of years ago that he sold the farm to his son Gordon, and bought the trailer house. The trailer is parked up close to the big farmhouse and Eddie lives there alone. It is parked right on the edge of the hill where you can look down the slope across the Marsh, over several cornfields and some low brush along a creek bottom and when the deer hunting season starts all Eddie has to do is sit by the window of his trailer and watch down the slope. He can see the deer come out of the Marsh, and if he feels like it he may take his rifle and try to get a buck. Sometimes the deer will come right up near the trailer and into the garden. Eddie likes that. Although he has been a great trapper and hunter, he feels a very close kinship for all the wildlife. He believes that we have to take special care with the wild things, to assure that they do survive and that they add an element to our lives that we can

get nowhere else. The wild things, Eddie thinks, are brothers of ours. We should treat them like brothers . . . with respect, and cherish them. The great life he has lived in and around the Marsh has been made real and alluring chiefly because of the wild things. Without them, Eddie says, we are a lost people. We will never know ourselves, or what we ought to do. He thinks that the ancient Indians must have understood such things a lot better than we do. Look at the forms they chose for their ceremonial mounds: beaver, panther, turtle, birds . . . all wild things. All nature. All sacred.

Evening On The Marsh

138

We start the truck and back out of Eddie Lehner's yard. We see him watching us from the window of the trailer. He waves. We wave. Lechner gets the truck into first gear and we pull out of the snowed drive into the country road. The gear has a peculiar, rather loud hum. I shift my recording equipment on the seat. We ride a quarter mile in silence. Just the gear noise and the truck rumble. Finally, I say:

Gard: What an interesting fellow.

Lechner: Yep. Some man!

Gard: Just getting warmed up. Gosh, what a trapper, though. Must be one of the greatest trappers in the state, huh?

Lechner: Rat trapper, I'd say he was. Yeah. He's quite a guy.

Gard: He's got a pretty big farm in here, hasn't he?

Lechner: The farm isn't so big. But he sharecrops. For the Federal Marsh. Or his son does. Now this road we're going up in here is a dead end. You go to the left and you hit a dead end again about a quarter of a mile down. That's what they used to call Smut's Point up there. The reason why it's named Smut's Point is that once there was an Indian village up there. It was said to have smelled so bad that the local people got to calling the point "Smut's." Well, they couldn't afford much sanitation or anything like that, and I bet most of the white settlements smelled the same way.

Gard: If Eddie got three dollars a hide during the World War II years, and if he took three or four thousand rats every year he made some real money. And I guess he has stories enough to keep going for a month.

Lechner: He does. That's right. He spent more time in the Marsh than he ever did on his farm.

Marsh Mink

Gard: And probably made more money trapping, too, didn't he? It seems as though they might deplete the muskrat population. How can you trap so many in a season, and have any left?

Lechner: It's impossible to extinguish them. Muskrats are cyclic. They're up and they're down. When they are up you have to trap them. They would cut the vegetation to almost nothing. They get Errington's Disease that affects muskrats if they become over-populated. Since muskrats will breed beyond the capacity of the food to support their population . . . they just die off.

We pass a large Marsh hay area, where, Lechner says, is the location Pete Feucht did his hay baling. They don't cut the hay so much any more. There isn't the same market for it to pack those beer cases over in the Milwaukee breweries. Some farmers do come and cut the hay though . . . We travel a while in silence and at length I remark that there seems to be an awful lot of water on the Marsh, and in winter, when the ice is on and snow covering it, you can see what big expanses there are where no vegetation is growing.

Gard: A lot of water on the Marsh.

Lechner: You bet there is. And this draw-down they are figuring on doing this spring, to get rid of the rough fish, you know, they will keep the water down all summer and a lot of the Marsh will dry up. A lot of the vegetation will come back.

Lechner stops the truck and points away out into the Marsh where I can see a dark line crossing.

Lechner: That line you see out there is called Lehner's Ditch. And see that big frame farmhouse over yonder? That's where Eddie used to live before he had to sell out to the Federal people.

Gard: That's where Eddie was born?

Lechner: That's where Eddie was born and raised.

Gard: Now why would he ever leave a place like that? It's right out in the Marsh. It would have been perfect for Eddie.

Lechner: He had to leave. Had to sell to the Government. Can't you imagine living out there in the fall when the geese are comin', and the whole sky full of them right above your house, and in the early morning you could hear the yellowhead blackbirds calling? Wild! We don't know what it is like living in the middle of wild things until we could live in a place like Eddie had. Can you blame him for being so sold on the life of the Marsh and all it means?

As we drive we are both silent. I am thinking that it's really a shame when the private landowners who have established the kind of homes they want suddenly have had to encounter a Government, no matter how benevolent or well-meaning, which forces them to get out . . . to leave the things they love . . . even if they know that in the long run the changes will do fine things for the country; it isn't ever going to be the same again as it was when grandfather came in and settled away off from anybody else, just because he liked being alone, away, in the middle of a nature he could appreciate without words . . .

140

Lechner: Eddie Lehner, as you could see, is easy goin'. He didn't make too much fuss about sellin' to the Government. Some folks did. Now right here where those trees are, that was the Oscar Starr farm. The old gentleman is still alive. He lives up there with his son; and back over in there, up to the dike and all the way up to the north end, there is a parcel of land, up north of Town Line Ditch, it was owned by Bernard Schabel. Good old Bernard, he's dead long ago; his son inherited this eighty. His name is Joe Schabel. Joe is a man now away up in his seventies. He owned that. And I used to trap that farm. And this was some of the land that the Federal Refuge had to buy. Nick Carter was a Federal land buyer, and Carter came to me, because I was a real good friend of the Schabels, and says, Ed, how can I buy that land? He says, I don't want to serve them with condemnation procedures. We try to get along without doing that. It makes hard feelings. Well, I says to Carter, let me talk to them. So I went to Mr. Schabel and I says, Joe, they are going to force you to sell this land anyway. And maybe I have a way to make a little something extra on it. Sell it to the Government with the trapping rights for one year, reserved. Well, that's what he did, and I got the right to trap that land. That's how I was the last trapper on the Federal Marsh! And Joe and I shared the money we made; and the Government got the land without a big fuss!

Gard: If a fellow just trapped in season, could he make enough to live on? The way things are now?

Lechner: Not in this day and age. Because you see, synthetics have come in, and there's all the great feeling against using real animal fur for garments, now. The price just isn't there any more. And there's all these imports, these furs from abroad.

We pause on higher ground where we can look north

The Mystery Beckons

over a large expanse of the Marsh. It looks lonely in its winter aspect and very desolate but still, there is that wildness. One thing about the Marsh, if you go into it from the east side, you've always got to come out again. Because you always run into a dead end.

There just aren't any roads going clear across the Marsh except Highway 49, away up at the north end, and the Federal dike, it goes clear across, too, but it isn't open for auto traffic.

Gard: What was the Government's reason for building

Boy At The Neda Mine: A View Of Nature

a dike across the Marsh? Was it just to separate the State and Federal portions?

Lechner: It was just that they wanted a little more water on the Marsh. They figured it would be better for ducks, and other waterfowl, and would make a better refuge for wildlife. You see that ditch down yonder, Bob? You see that road goin' away up into the Marsh?

Gard: Yes.

Lechner: That's the dike. That goes clean across. About halfway it doglegs down to the south, then it goes west again.

In the old days the roads in and around the Marsh were never like they are today. It was all shrubbery and brush, and none of the roads were gravelled, even. The bottom sort of dropped out of the roads in the spring, or after heavy rains, and it was hard, sometimes even dangerous traveling. But the old-timers knew the roads and where the bad holes were.

Ed Lechner said that just down from the old Feucht place, not far from the old road, was where he saw the last covey of quail he ever heard about being in the Marsh. And just north of there, where I could only faintly see some tall pines, now, that was where Ed said the last flock of prairie chicken used to roost. The prairie chicken were very plentiful in the old days. So, so many, Ed said, and he saw these last flocks in WPA days back in the 1930s when they were working in the Marsh as part of a CCC project. Up back of where Ed said he saw the last flock of prairie chicken, and a little to the west of it, was where Bill Schreiber lived. And he pointed out the hill that was about half cut away . . . when the Federal land movers, after they had bought Bill's farm and had moved him out, despoiled the hill to get earth for the Federal dike.

Lechner: You see this little woods down here? From that you move straight over to the west. That's where Schreiber's house was. The Government tore that down, of course.

Gard: Beautiful place for a house, and I don't blame Mr. Schreiber for being some angry, when he had to move.

Lechner: Oh, I don't neither. This was a paradise. This was their life. You can kind of see that hill now, how it's cut down there. From the dike to the south is State of Wisconsin owned and to the north it's Federal owned.

Gard: Does anybody fish in the Marsh any more?

Lechner: Oh yes. But the carp have ruined it. That's why we're going to treat the waters this summer. Then fishing will be real good again.

We passed a very neat farm, tucked back under a hill, with the Marsh spread out all behind it. This, Ed said, was owned by a friend of his, Arnold Luebke, who has

five farms. Hand-me-downs in his family. And Earle's mother is a sister of Bill Schreiber. So a lot of intermixing of families does and has gone on in and around the Marsh.

Lechner: They used to cut a lot of Marsh hay out of here when I was a boy. Team upon team, cutting and hauling hay all day long. Just like old Pete Feucht told. Years ago, a professor from the University introduced alfalfa, and when this happened the farmers forgot about Marsh hay. The only hay that the farmers raised after that was Timothy hay. This was for horses only.

We passed a road Ed called "Northern" road. And the reason they called it that was because, in the Rock River, which bends around right along there, they used to catch northern pike, twelve, sixteen, twenty pounders, all the time. It was the most terrific northern pike fishing in the whole state, until the river got polluted and the great fish went away, or died out.

Back up above, where we were easing along in Ed's truck on a very lonely road near the river, was the old Krueger place. And the old pump, Ed pointed out, is still up there. The Kruegers were really back in. I remembered how Mrs. Krueger had told us how shut in they were.

Near the Marsh are several small towns. They sit in the open spaces, quite isolated it seems to me, and the character of them is like the same image of all small towns that were once important places and now shelter mostly memories.

Lechner: This is the town of Knowles. It was spelled K-n-o-w-l-e-s but it was named that because there were so many knolls or hummocks in the vicinity. At least that's what the local folks say. Folks around the Marsh, and in these little towns, or the farmers living nearby . . . some years they had it good. But many of them had to do a lot of different things to make a living. Sometimes they were trappers, or farmers, or some started businesses.

The shores of the Marsh have always been pretty good farmland. You couldn't have put a house out on the Marsh itself if you had wanted to. Would have sunk right down, unless, that is, you got on one of the islands. There were and are several of those. Stony, Steamboat, Goose, Four-Mile, Cotton, Misling . . . You couldn't have farmed the Marsh either, because in the older days they couldn't ditch it or they couldn't tile it. Though they tried. You can farm Marsh land if you ditch it and tile it, but this costs a fortune. The Federal Government has had the money to do this to a small amount of the Marsh lands lately, and they raise some corn in there for the wild geese. But the State doesn't have any farmland right in the Marsh itself. They sharecrop with the farmers who live around the edges . . .

Wonderful and beautiful wetlands, the cold waters of spring have broken free in the river and in the ditches. In the channel of the river is the image of deep white clouds, and a spring stillness is over the whole Marsh, trembling until the splash of the landing duck breaks it; or the wild song of small birds begin . . .

Then in the morning, gliding with no sound, for the splash of a paddle carelessly used would break the spell, comes the canoe, and the man and the boy in it have on their faces the quiet and joy of an early Marsh morning.

Far away, to the south, is the ledge of high limestone where the watchful men of older races saw the light move across the wetlands. Now, far away it seems, for my mind is wandering in distant ages,

come voices of children; for there are modern campers now on the ledge. The wonders of ages past with still faint memories of a tall blue glacier, and in the fields to the west are the stones of a glacial drift. There is purity in the wetlands in spring; solitude is purity. State and Federal Governments have tried to preserve the solitude and recapture the purity.

By 1946, the State of Wisconsin had purchased the southern one-third of the Horicon Marsh. During the 1940s, the U. S. Fish and Wildlife Service purchased the northern two-thirds and constituted it a National Wildlife Refuge.

The renovation of the Marsh started with the construction of the new State dam at Horicon in the early 1930s. Legal difficulties concerning the opening and closing of the dam followed until in 1943, when final control for its operation was vested in the Conservation Department, now called the Department of Natural Resources. The dam, of course, stopped the uncontrolled drainage through the old main ditch. However, if the dam has held back enough water to wet all the desired areas, some portions would have become lake rather than marsh. So, after setting the water level at the most advantageous point, other means were found to increase the Marsh wetlands.

The wetlands on the Federal portion of the Marsh were increased between the years 1948 and 1951 by the construction of the dike which crosses the Marsh from east to west. This dike, located a short distance north of the State-owned portion, holds back enough water to flood an additional 12,000 acres on the Federal area.

So for many years, since the restoration of the Great Marsh, the struggle has continued to make the Marsh come back into its own, and to realize something of its former state of plenty. The Marsh is ordinarily covered by an average of two feet of water, and contains semi-dry scattered blocks of uplands. Not all the Marsh areas are swampy. Dry upland areas are needed for geese and other game birds and animals. In some upland areas, shrubs and trees have been planted to provide food and cover for game. Other upland areas are kept as fields where crops are grown to provide food patches. The State farms 300 acres and the Federal Government farms 1,912 acres, much of it on a sharecrop basis. Crop acreages on the Horicon National Wildlife Refuge have been increased in recent years by expansion of peat land units farmed by Refuge personnel. These units now amount to about 628 acres. Although these food patches are planted primarily for geese feeding, other species — ducks, pheasants and deer also utilize them.

Until 1967, the harvesting of muskrats was done through a share-trapping program in which the trappers and the Federal and State Governments shared in the proceeds from the sale of the pelts. However, in 1967, all of the fur was given to the trappers, because of the downward trend in skin prices. Permits are still required though and trappers are assigned specific areas in which to trap. The purpose of the trapping program is to keep the muskrat population within desired limits. The muskrat plays an important role in Marsh management as it opens up dense stands of aquatic vegetation through food-cutting and house-building activities and creates small, open-water areas that become loafing and feeding sites for ducks.

On the Wisconsin State portion of the Marsh, several dry areas have been converted into wetlands attractive to ducks and muskrats by the construction of level ditches. These flat-bottomed ditches are not meant to drain or divert water, but to hold water. Unlike the feeder ditches of the drainage era, the level ditches are closed on both ends. To date, sixteen level ditches have been dug on the east side of the State Marsh, adding ten miles of new water area. Drier portions of the Marsh also have been made attractive to waterfowl by the construction of dikes and the use of lift pumps for flooding. These areas are called "impoundments" as water pumped into them is retained or impounded by the dikes. When crops, such as buckwheat or Japanese millet, are planted prior to the flooding of the impoundments, they are made even more attractive to ducks. There are five such areas, totaling 746 acres on the Federal area, and four on the State area, totaling 445 acres.

Controlled burning of dead vegetation has been practiced on the Marsh since 1946. The amount of burning varies from several hundred acres to several thousand acres, depending upon weather and water conditions. There are two reasons for this practice. The removal of dead bulrushes, cattails and sedge exposes the soil and enhances the growth of new vegetation in the spring, thereby providing favored resting and feeding areas for ducks and geese. Secondly, the repeated burning retards the buildup of the Marsh floor and the consequent changes in the vegetation to less desirable forms.

The State Marsh management has maintained fish traps on the main channels of the Marsh for carp removal since 1943. Two such traps are presently operated each winter when the fish move into the main channels seeking water with a higher oxygen content. Up to a million pounds of carp are removed annually through this means. As the carp is destructive to aquatic vegetation, it is desirable to keep its

Detergent Foam: Pollution Has Been A Problem

Warden Burrows Checks Fish Life, Here Bullheads In Jeopardy During Freezeout

numbers under control. Other more desirable fish, captured during these operations, are returned unharmed. In addition, an average of 1,000 adult northern pike and thousands of fry are stocked nearly every year in an attempt to offset annual winter losses of pike.

One of the most dynamic of the older state management men is Barney Wanie who took the State of Wisconsin examination to be a law enforcement officer on the Marsh in 1934. Now retired from service, Barney came to Horicon the ninth of October, in 1934, shortly after the dam had been built, to raise the level of the water on the Marsh again.

Farmers in the whole area were enraged that year. They did not want the flood waters to come back. They had opened agricultural fields and had been farming them for years, while the Marsh was drained, and now, suddenly, the fields would be wet again, and farming impossible along some of the edges of

Fish Life In Jeopardy

the Marsh. The farmers threatened to blow out the dam and Barney and another warden O. K. Johnson, spent twenty-four hours a day, every day, guarding it.

Methane gas and hydrogen sulphide gas have a very bad odor, and certain areas of the Marsh are underlain with these gases. When Barney first came to the Marsh in 1934, the odor of the gas wasn't bad, but now if a hole is chopped in the bog anywhere off the main river, the gas smell is overpowering.

Barney told me that he came to the Marsh to be in charge of rough fish control. And he built the first fish traps in the State of Wisconsin. They were built on the Marsh to catch carp. There was one down the river from Horicon, one on the east branch of the river, and one on the north branch near Waupun. Nineteen-forty-two had been a hard winter and many fish died. Barney was accused of killing off the fish. The German farmers were irate, because they wanted the carp for themselves. They saw a lot of dead fish and they blamed Barney for killing the fish with the traps. He knew that he hadn't killed them and he was able to show the farmers that it was the lack of oxygen when the ice froze so thick that killed the fish. The winterkill of the fish, though, didn't really get so bad until pollution of the waters began. When factory byproducts began seeping into the river was when pollution got bad. Barney suddenly knew that something dreadful was killing the fish in the Marsh. The gills of the fish were normally red, and they died in thousands; the gills turned white. While this is normal, Barney suspected that the fish had been poisoned.

Barney took water samples and sent them to the lab in Madison and in a couple of days the chemist called Barney and said, Barney, where's cyanide come from?

Barney put on a whirlwind search and he found out in a hell of a hurry where the poison was coming from . . . a pipe running into the Rock River at Waupun. Barney says now that in those days you couldn't tell the management of a factory like that factory in Waupun to cut that out . . . stop putting that poison in the river. They would have just told you to go to hell. But Barney went to the company and got them to view the dead fish, and because they respected him, they listened. They put a special machine in the factory, and when they dumped the cyanide out, they cleaned it in the machine. They found they actually could save money by not polluting the river.

The company had no more trouble until about seven years later when the machine broke down one day, and the men, not knowing, dumped the water out. They had a bad fish kill again.

There was another company that manufactures milk products. That company took the solids out of whey, and the by-products went into the river. There was a

Dipping For Carp

pipe that was sixteen inches through. And it ran day, after day, into the river. The waste created a solid mass of foam six feet high over that part of the river. Fish were killed by the thousands. The State took the operator of the factory to court.

Suddenly, one day, the whole case was stopped. The State men were simply told by somebody, no one was sure who, to lay off.

The deadly winterkill on the Marsh is caused by the decay of putrefactives on the bottom. In the bog on either side of the river, the fish can't live because of the methane gas, and lack of oxygen.

There is a famous yarn told by Emerson Hough, about bullheads in the Rock River that should be told here; because it shows, in an exaggerated way, no doubt, how the fish crowded up under the ice to search for oxygen. Hough said that:

I have received permission to tell the Kekoskee Fish story provided that I do so in a careful and temperate manner. I do not know how to obey this injunction than by giving it in words of Dr. Clark of Mayville from whom I first heard it.

These events happened before the war, said he. They are so singular and improbable that I always hesitate about telling the story. You will probably laugh at me and not believe me, yet every word of this is true.

The winter of 1860 was very cold. At that time a vast lake covered the whole ground where the Horicon Marsh now is. The lake was full of fish, and when the ice had frozen deep over every portion of the lake these fish became distressed for air. The Rock River, as you know it, is a lively stream here, and as you have noticed, it has a stretch of swift water just below the great dam at Kekoskee. The dam existed at the time of the story. You have looked with your own eyes upon the very spot where these startling incidents occurred.

The fish, unable to breathe in the half solid lake, crowded up the live channel of the Rock River, making for the hole which the swift water kept open in the ice below the Kekoskee dam. Most of these fish were bullheads. It is six miles from the lake up to the Kekoskee dam and the ice on the river was two feet thick, yet the whole bed of the river, forty yards wide, was, for a distance of six miles, so packed with bullheads that the heavy covering of tough ice in places rolled and tossed like waves of the sea, so desperate was the struggle of the horny host beneath it.

The first arrival of the run of fish at the open hole was marked by a geyser-like eruption of bullheads fifty feet across and about twelve feet high. The pressure of the fish behind was enormous. The fish could not get back in the water and so slid out on the ice, covering it in every direction for hundreds of yards to a depth varying from six inches to two feet.

The air was filled with a strange, low murmuring sound, which could be heard nearly a mile around. Old settlers say they never heard such a sound since. Dreading some unknown calamity they hastened to the spot, and there, as you may suppose, their dread was turned to joy.

Before noon of that day every team of the whole neighborhood was at the dam hauling bullheads. The amount of bullheads taken from the spot I hesitate to state for fear you will not believe it. No one believes

150

it. They always laugh at us when we tell the story and think we have gone crazy. In Wisconsin the term "Kekoskee man" is used to designate one who has a wheel in his head. No Kekoskee man has been believed on oath or admitted to the jury in Wisconsin since 1860. This unearned reputation has ruined the town. You see it as it is, silent, almost deserted, a few empty buildings standing in the town martyred into ruin by too strict adherence to the truth. For every word of this story is true.

If you will come with me about a mile out in the country, I will introduce you to the widow Schneider, now an old lady. The widow Schneider will tell you that on one morning she counted 900 wagon loads of bullheads on their way from the geyser below the dam. This was only one morning and the run lasted two weeks. Of course, this number of wagons only represented a part of those which passed, and this was on only one road of several leading out into the country.

The bullheads were shoveled into the wagons like potatoes, and the regular price was twenty-five cents a load — a nominal sum to cover the shoveling only. One man who shoveled there bought himself a farm in the vicinity with the money so earned.

The bullheads were hauled out in the country and used largely for manure. There is no richer land in Wisconsin than this has been since 1860, and all the farmers fed the bullheads to their hogs, and for two years after that you could not get a decent piece of pork in this part of the state. It was all fishy. The hogs all took naturally to worms and liver after that, and some of them evinced rudimentary gills behind their ears. Oh! I don't blame you for doubting this. They all do.

There was a ford in the road at this point of the river, but the wagons could not get into the water. After the first eruption of bullheads had subsided, planks were laid across on the living pontoon bridge of fish, and on these the teams crossed.

Even after the run had subsided very much, dogs and children were known to run across the open hole on the backs of bullheads. Still later in the run, after the fish had thinned out a great deal, a man well known in this community, Julius Cornell, slipped from the ice and fell into the hole. He could not get into the water for the fish. You smile at this. I do not blame you. We are used to it. No one ever believes this story.

After the bullheads thinned out so you could get a spear through them as they lay in a matted layer, it was discovered that there were layers of bass and pickerel lower down in the water, and much sport was had with these later on.

Of course, everybody that winter lived on bullheads, and they were used in many ways. As I have said, the farmers fed them to their hogs. We had a lazy sort of an expressman here named Brush, and he owned a fallen down old horse which dated back to the Mexican War and was called Santa Anna. Brush insisted that he was too poor to buy Santa Anna hay, and so he fed him bullheads all winter, and that was everything the horse had to eat for six months. Oh, laugh if you want to, we're used to it. But I'll have you out and show you Santa Anna, a good healthy sort of a horse today. Brush had moved to Bayfield, but you ask any citizen of that town if Santa Anna didn't live on bullheads, and if he don't tell you just what I have, I'll retract the whole story. You needn't think that I am talking to you out of my head. These things are all facts, and you can get all the proof you want. You just go out alone, don't take me along, but just stop any citizen of Mayville you meet and ask him how about the Kekoskee bullheads. That's all I ask you to do. You just sift this story and see if you don't find it is true.

We did sift the story and we did find it true. That is the singular thing about the story, and that is why I call it the most remarkable story I ever heard. The facts themselves are not beyond the range of imagination, but to have a whole community rise up and testify to their truth — that proves that imagination had nothing to do with it, and that the facts are facts, pure and simple. Ordinarily one man tells a fish story, here 200 tell it and tell it just the same. The evidence is legal, convincing, overwhelming. In the total, it makes up the grandest fish story ever told. I tell it here, but it is nothing. No one man can tell it. To hear it aright you must go to Mayville. There the whole town will tell you this story. You dare not, can not doubt it. You will believe and you will feel as we did here that the entire chain of circumstances in this case constitutes the most remarkable incident in the history of a lifetime.

Marking Geese For Tracing

Winter Is A Hard Time For Fish

We met F. M. Lawrence, leading attorney of the town, and asked him carelessly about the story. Mr. Lawrence was alone and not in reach of Dr. Clark. "Oh, about those Kekoskee bullheads," said he. "Yes, that was a great time. The widow Schneider, out there a mile on the road, counted 900 loads go by in one morning. A dog could run across the open hole on the backs of fish. Julius Cornell fell off the ice and couldn't get in the water for the fish. The farmers fed bullheads to their hogs, and one fellow here in town, named Brush, kept his horse all winter on bullheads. He had them stacked up out in the yard. The horse's name was Santa Anna. Want to see him? Come along with me and I'll show him to you. A good many people don't believe this, but I'll show you the horse."

We happened across Ed Sauerhering, who owns the drug store, and dropped a hint that we had heard something of the story. He was alone and couldn't see Dr. Clark and Mr. Lawrence. "Oh," he said, "those bullheads? Yes, there was an awful lot of them. Julius Cornell slipped off the ice and fell into the hole, but he couldn't get into the water for the fish. That was in the cold winter of 1860. The widow Schneider counted 900 loads of bullheads go by her home one morning. Yes, the farmers fed the bullheads to their hogs. Old man Brush, an expressman here, kept his horse, Santa Anna, all winter on bullheads. He said he couldn't afford hay and had plenty of bullheads. Do you want to see Santa Anna? Come along and I'll show him to you."

At the hotel we met the landlord, an old German, and we had a word or so about the story. "Oh," said he, "Dose bullets? My, dat was an offul dime alretty. Yulius Gornell, he fall the ice off, and he couldn't get into the wasser bei der vishes yet. Everybody feed dose bullets bei der hogs and I gets me so tired of bullets I been sick of life. Old Prush, the egsbressman, he feed bullets all vinter bei his horse, old Santy Anny alretty. Yes, dot Santa Anny don't eat nodings else ven der bullets last. If you vand see dot Santa Anny you come along, and I shows him to you. Der vitter Schneider, dot lives one mile on der road out, she hat doch nine hundret vagon loats of bullets one einzigen morgen gesehen. I show you old Santy Anny und der vitter Schneider. Ya, der is no doubt apoud dose bullet, not by along vays yet."

That evening we sat in the hotel office, a kind of a town meeting ground. One by one as the different citizens came in at the door, they were without previous hint, without possibility of collusion, asked to stand and deliver the Kekoskee bullhead story. One by one they did so without hesitation and with absolute unanimity on the points of Julius Cornell, the widow Schneider, and old Santa Anna, as well as on minor details.

This is what I claim for the Kekoskee fish story. It is not a plain lie and it is not unbridled imagination, but the best possible exposition of facts which do not admit of doubt. The evidence is there and any man can go to hear it. If he does, he will doubt no more, believe as I do, and as I hope all readers of this do. Moreover, I myself have seen the spot where the hole in the ice was. I have seen the house of the widow Schneider and I have seen Santa Anna. I should dislike to have this story meet the ridicule with which I treated it before I had become convinced of its entire truthfulness, and doubting allusions to it I shall treat as personal aspersions. If the proof offered here is not sufficient, there is plenty more in Mayville.

Note: Dr. Clark claimed that Mr. Hough omitted one very important phase of the story, which is, that the pressure of those bullheads finally became so great as to actually reverse the operation of the mill wheel, causing the wheel to revolve against the water flow. He says, however, it's so long ago though and so many of the old settlers who remembered and would bear out this statement have passed away, that no doubt most people will doubt the truth of this last statement too.

Nevertheless, and in further substantiation of all the foregoing, be it remembered that after this tale was given to the public, F. M. Lawrence was elected Judge of the County, and Ed Sauerhering was sent to Congress, from the district in which Kekoskee is located, and Kekoskee residents are once more believed under oath, and also permited to serve on juries.

So during the winter, the fish concentrate in the main river. The decaying vegetable growths consume the oxygen beneath the ice, and when the oxygen is gone, the fish die.

It was the greed of men, though, that caused some of the great winterkills. Where there was an air hole the fish gathered in great numbers. Men from far away came with wagons, and they would use a wooden minnow, or a tin minnow, and wave it back and forth in the clear water. When a fish would come the men speared it, and tossed it into the wagon. Fish were so plentiful that farmers would return home with sleighs filled with Rock River fish. This went on through the cold months of the winter. There was no one to stop them. Bullheads were often just scooped out of the water with nets or shovels and wagons or sleighs filled.

One winter Barney set a large fish trap in the Rock River, and there were thousands of carp coming up the river. He trapped 300,000 pounds of carp, and had them all in the trap. They used up the oxygen so fast that Barney knew he would have to supply it. He installed a pump under the ice and purified the water with the one pump and sprayed water onto planks and it ran down into the river. He kept the fish alive for a month with the pump.

Barney took 900,000 pounds of carp out of the river that winter, and he knew, from the way the fish hurried from the lake to the river, where the water flowed, and was deeper, that they were running from the threat of no oxygen. He was able to trap them because of their fear. And he knew, too, that man was causing the demise of the fish in the Marsh and river, because he was able to measure the increasing amount of silt and debris running from the surrounding farms, and causing growth in the Marsh, and then this growth decayed, and as the water became shallower, the danger of the winterkill became greater.

President Grant signed the American Fish Commission into existence in February, 1871. For a long while the Commission's greatest achievement was said to be their introduction of carp into the waters of the United States. The carp was a favored food fish over a great part of Europe and the Commission introduced it to provide a cheap source of food for rural people in the south and midwest. Rearing ponds for the carp were even constructed on the grounds of the Washington Monument and other locations in Washington. The Commission introduced the carp into every body of water they could possibly reach. However, sportsmen groups soon grew suspicious of the practise. Clouded water that once was pure, and the reduction of game fish in many areas grew. The Commission was soon accused of launching into a large program without adequate study. But by then it was too late. The carp were here, and they are still here and their number is rapidly increasing.

The carp, of course, rooted out the plant life, and when there is no plant life to give out oxygen, this menace of the carp added just one more hazard. The carp are also a congregating fish. They like to be together; when thousands of pounds of carp get together they use the oxygen, and often under winter conditions, they too die. This tendency to congregate is one of the chief reasons why carp are so easy to trap. The carp, however, is a very hardy fish and can survive under difficult conditions . . . more than any other fish, perhaps except pike.

Carp were only regulated at first by taking the largest ones . . . commercial fishermen didn't bother with small carp. Along with the carp, in the waters of the Marsh, are the bullheads. The meat of the bullhead is very good . . . many persons relish the fish, and consider it one of the best eating freshwater fishes. The bullhead is a grazer. He will eat vegetation from the top. The bullheads will eat the vegetation right out of a lake.

In the days before the carp were put into the waters in the 1870s, mostly by the local German settlers who liked to eat them, there were plenty of black bass in the Marsh waters. But in those days everybody had all the fish they wanted. Even small boys, wading in the rapids at Horicon, could catch sizable fish with their bare hands. Northern pike, black bass, walleyed pike. The Marsh was known far and wide for the excellent fishing. People often drove teams clear from Madison, just to fish. And the fish grew large.

No matter how many fish were taken from the Rock River and the Marsh waters, there seemed al-

Helicopters Used In River Cleanup

ways to be even more the next year. The fish had excellent food and they grew fast; the northern pike would run sometimes thirty pounds or better, and the great fishing lasted on the Marsh until the pollution killed it.

The large attempts to take out the rough fish and to make the waters more habitable for game fish have been a very good idea, Barney thinks, but he wonders what the end result will be when nothing much is done to prevent the pollution and the erosion. Those are still the greatest enemies of the fish in the Marsh area.

The whole part of the river from the city of Mayville down to Kekoskee was an excellent location for largemouth bass. There were stones and rapids by the bridge where an old wool factory formerly was located; but when the stream reached the Marsh it became a different story. The mud collected. It became bog. The fish tended to leave the Marsh and seek the clearer water of the river.

Beginning in 1969, the State of Wisconsin began a "draw-down" of the Marsh waters, with the purpose of getting rid of the carp. I visited with Bill Selbig, scientist in charge of the first renewal or "draw-down" project. He said that the carp muddies the waters and roots up the vegetation — causes such muddy conditions that there is little oxygen and

hardly any photosynthesis. The carp are a very sturdy fish and seem to survive the winterkill better than game fish, and within thirty years the whole picture of fish in the waters of the Marsh changed from pike, walleye, black bass and other game fish to almost complete dominance by the carp. There are a few game fish left, and within the 30,000 acres of flooded land, there is some reproduction of game fish, but the young fish die every winter.

Now the conservation men will change the pattern. The carp must go. There will be a mighty draw-down. The waters will be let out of the Marsh, the dam opened at Horicon, and below, at Hustisford, and little by little the Marsh waters will recede until perhaps by midsummer 1972, the Marsh will be empty of water except in the Rock River channel and in some of the ditches. If it is a dry summer the peat will dry down, and for a while the Marsh will resume the appearance it had in those other days when the drainage men drew off the waters. Only this time the purpose will be to change the patterns of fish life, so that the great days of the game fish may return.

Planning for the draw-down started in 1969, and in May of 1971, the action phase was started with Bill Selbig in charge. In August, 1971, the men put fish toxins into the east branch of the Rock River and most of the carp were removed. Now the entire Marsh will be treated. Of course, in time, the carp may come back again. There is nothing to assure that they will not, though there will now be an electrical fish barrier at the Hustisford dam, far south, to prevent the northward spring migration. And too, the conservation men now have a method of preventing the winter-kill, or most of it, through pumps and air guns that supply oxygen beneath the ice.

Within days after the carp are removed the water becomes crystal clear. In shallow, warm waters the carp are chiefly responsible for keeping the mud and muck of the bottom in suspension.

Until fairly recently only small lakes of two or three hundred acres could be successfully treated with the fish toxins. The large waterways such as the Rock River and the larger lakes couldn't be helped, though it had been known for years that they needed attention. The use of helicopters and fixed-wing aircraft, new formulations of chemicals, and new techniques of spreading the toxins has enabled the contemporary conservationists to use the toxins wisely and effectively.

If there is a flowing stream coming into a lake, or into the Marsh, the men used "drip stations." They put the chemical into fifty-gallon drums and it dripped through the waters, ten parts per billion. The drip stations were spaced every half mile, or every mile, depending on the ease of access. And this method was followed for the rivers and the streams. If there was a lake to be treated or a standing body of water such as a drainage ditch, then aircraft were used to spread the toxin . . . helicopters or fixed-wing aircraft with sprays distributed it.

The amount of poison is calculated according to the depth and dimensions of the body of water. Antimycin is the chemical that is favored today. It is organic in nature and is not a hard pesticide. It degrades very rapidly, and within three days it has vanished completely from the water. Fish can be restocked immediately. The chemical has complete Food and Drug Administration approval. It is, however, extremely expensive.

There is another chemical having the same approval. It is called Rotenone, also organic, derived from the *Derris* root, from South America, where the natives have used it for centuries to stun fish. It is much cheaper than Antimycin . . . costing about three dollars a gallon. The fish, however, can sense the chemical in the water. The carp, so think the game men, is probably the most intelligent "critter with fins" and the carp, when they sense the chemical, will sometimes find a way to escape. A hundred percent success with Rotenone is very rare. Antimycin cannot be detected by the fish.

DPN Representative Helps Boy Scouts "Plant" Wood Duck Houses They Constructed As A Project For Horicon Marsh

Both chemicals are effective on all scale fish, and all fish have some tolerance to them. If the carp are killed, most likely all the fish in the lake or stream will also be killed . . . a complete blot-out, in other words, and a complete start-over.

Catfish, however, have a very high resistance to Antimycin and are relatively secure.

In the Marsh, where the concentration of carp has grown so vast, the collection of the dead fish is a serious problem. It takes more organization and effort to accomplish the dead fish pickup, than the actual treatment, as far as "on the spot" effort is concerned. For the pickup of the dead fish more manpower is needed than the conservation men can supply from among themselves, and citizen volunteers must help. Fortunately, many persons interested in conservation matters do offer to assist.

One pint of Antimycin costs the project $21.50, and a pint will treat, roughly, about four acres of water, one foot deep.

The conservation men have made, up to date, very exact hydrological maps of the entire Marsh area so that precise amounts of chemical may be used. Everyone is terribly interested to see what the Marsh will look like when the water is drawn.

Somewhere around Labor Day, 1972, the dams at Horicon and Hustisford will be closed again, and slowly the waters will return to the Marsh, and with the new, clear waters, new fish will be stocked, and hopefully, the grand fishing days of the past will return.

The men expect that there will be many tons of debris on the Marsh shorelines . . . the carelessness of hunters, trappers, tourists, campers of many, many years will be exposed, picked up and put away.

The cost of the draw-down isn't as much as one might think. It costs only $210.00 per mile to treat a stream system such as the Rock River. The boys at the Marsh think that's a cheap price if the rough fish can really be got rid of.

Keeping the rough fish under control has been hard work. It is work that has occupied the time and skill of many good men. Now, with the draw-down to remove the rough fish from Marsh and river, the job, for a while at least, may not be so intense. But the larger ones still fight their way upstream.

158

I have developed a deep emotional reaction to the Great Marsh and its people . . . The history, so filled with lore of the Indians, and the struggles of the white settlers, and so rife with the crucial changes that dams and drainage and their resulting legal battles have brought to the Marsh . . . all these things feed an emotional channel in me that always leads to wonder of what is now to transpire . . . after all the struggle, what now?

The fact is that the Great Marsh like every other marsh if left alone, would strive to fulfill its cycle and to become, out of its transition from wet to eventual dry, a wetland no longer. But now that the Marsh is a refuge, and one that is increasingly used by migratory birds, the wetlands are necessary to the bird life, and have to be maintained.

It seems to me now that the climax portion of the story of the Great Marsh is the wildlife rather than the human history. The human story, in a sense is almost finished. The purchase of the land by government, scientific care in directing what must be done to protect wildlife and the whole ecology. Dedicated workers. Federal and State funds have combined to bring the Marsh back to a fruitful life. If anything, the struggle has been too successful. For example, the greatest wild geese congregation in America takes place at the Marsh now. Ducks were the big thing to the older hunters. There just weren't any wild geese to speak of, until in the 1940s. But now, as part of all that has happened at the Marsh, it is the Canada goose who takes center stage. These birds furnish the most dramatic spectacle in the Middle West in the fall and spring. In a way it is as though they have come home to reap the harvest of the struggle that has, really, been in their behalf. They are the living drama, and they complete a scene that, without them, is interesting, beautiful, and unique, but incomplete. For the Great Marsh without the wildlife, so symbolical of its character, and of the character and culture of the old Indian peoples, is only wetlands. When the birds are there it is living drama on a scale never dreamed by ancient or modern peoples, perhaps.

It is impossible for any person to imagine a quarter of a million Canada geese all more or less in one small area, resting, feeding, flying, talking to each other or to the sky, or whomever it is that wild geese address themselves to. It has to be seen to be believed. It is a sight so wild, free and so precious that even in its very preservation, the men entrusted to preserve it are deeply worried . . . the whole thing has been almost too successful . . . the results of the conservation programs almost too well accepted by the birds. The problems are now. Every day. Mostly, especially in migration time, the problems are urgent.

The Canada goose, above all other migratory birds, seems to symbolize, for the common man, often weary of daily routines and frustrated, perhaps, by thwarted ambitions, a sense of freedom and release. To stand quite alone in the open country and to see, of a day in October, very high, a great V of Canada geese heading south; or to watch the same sight in spring and to hear, faintly, the drift of sound that comes from the sky is a thrill like no other. The eyes shut and the mind wanders free, perhaps with the flying birds, heading off somewhere wild and unfettered. It is what a whole lot of folks feel when the migration comes through.

The Canada geese that stop at Horicon come there from their nesting grounds on Hudson Bay and west of James Bay. At their great speed, which, with a tail wind, can reach about seventy miles an hour, it takes roughly twelve flying hours for them to come from their nesting places to Horicon. Canada geese have been seen flying 9,000 feet high. They generally fly about the same route each season, and they are a part of what is called the Mississippi Flyway, which means that they generally follow the west shore of Lake Michigan to Horicon Marsh, then west and south between the Fox and Rock Rivers and on south to three refuges in southern Illinois, located near Carbondale, Ware and Cairo. Now about a half-million Canada geese use this flyway.

The Canada goose has always been a sign of the end of a hard, grinding winter, of promise of the burst of buds, and the melt of ice. They promise that the muskeg will soon come to life; that the bleak, snow-covered spaces of ice will turn to blue water, and that, in the marshy areas the cattails bring up green shoots among the brown seed hulls and rattling, dried reeds of last year. Traditionally, in the hard winter sections of Hudson Bay and James Bay, where the geese have returned to nest, the Indians have awaited their return with the impatience of humans who hunger for the spring and for new food. The Cree Indians of the lowlands make plans to feast on a share of the birds fattened at Horicon on their way north, for the geese have paused to feed and rest and have arrived in good flesh.

The Indians have, during the winter, made up many stories about prowess as goose hunters.

Old LaBoutille, ninety years old, likes to tell about the year game departed from the Saint James Bay area: the squaws foraged in vain for berries, and the children, too weak to play, cried themselves into what

was perhaps to be their last sleep. I was preparing to join my forefathers beyond the sky's horizon.

While thinking I was looking on the waters for the last time, I saw far away, but approaching me, many birds. They came nearer and I realized I was looking at thousands of geese! But alas, they did not attempt to approach the shore but alighted far out on the waters and commenced to feed. The sight of them kindled a great idea and stirred me into activity. I rushed back to my camp and shouted to everyone to prepare many cords of "babiche," each with a running noose at one end. While this was being done, I procured a whiplike sapling. The cords were many but light in weight, and I fastened them as well as the stick to my body, rushed toward the lake, and diving in, swam with all my might toward the flock of feeding geese. As I neared them, I sank beneath the unsuspecting birds. I unfastened the cords and with deft fingers slipped a noose around each pair of legs within reach. Moving this way and that, I secured many more until at last, my supply of "babiche" exhausted, I had captured as many geese as would form a feast for half the Indian people of the North.

Then up I popped in the midst of the still feeding birds yelling like one possessed and with my club stirred the startled geese into wildest activity. With a thunderous whirring of wings, they rose clear of the water, each one in doing so, tightening the noose around its legs. Up and still up they went, drawing me behind them clear out of the water and far into the air.

With cunning born of many weeks of hunger, I steered the flying birds toward my camp many miles distant. As we approached I gradually hauled in on my lines and as goose after goose came within reach, I would deal it a great whack with my stick, at which it would fall inert to the extent of its cord, thereby acting as a brake on the progress of the remainder. Then, as my lodge surrounded by astonished upturned faces loomed into view, I pulled in desperately on the cords securing the remainder of my winged steeds and smote right and left with my cudgel. Each blow lessened the speed of my flight, and as the supporting geese decreased in number, I gradually descended until with the last few struggling birds, I gently landed on the earth before my teepee.

From Parmenton Decorah, heir to ancient Winnebago folklore, I heard the story of how geese at the Horicon Marsh got red eyes.

Wauk-jong-ka-ga was the folk hero of foolishness of the Winnebago people. There were other heroes who represented human traits, but Wauk-jong-ka-ga embodied all that was foolish. One day he was walking beside the lake at the Horicon Marsh and he saw a large flock of geese floating around on the water. The

Roger Little Eagle Looks Back

The Swans, Early Arrivals At Kekoskee

geese saw the hero too, and they said to each other: look, Wauk-jong-ka-ga is coming to see us.

The geese swam over to the shore and when they got close to Wauk-jong-ka-ga they said: Hey, what have you got in that bag you are carrying on your back?

Why, he said, this bag is filled with songs.

Now that's great, cried the geese, why don't you sing some for us?

Why I would be glad to, said Wauk-jong-ka-ga. Just build me a wigwam with one door facing the lake. Then I will sing you all the songs that are in this bag.

The geese said they would do that, so they built Wauk-jong-ka-ga a wigwam with one door toward the lake. When it was ready he said, now everybody go inside and we will have some songs, and you can dance to my singing.

The geese replied, sure. We'll do that. So they went into the lodge single file and started dancing around the long lodge.

Wauk-jong-ka-ga started singing a song. And the song was that if you open your eyes while you are dancing your eyes will be red. If you open your eyes your eyes will turn red, and he kept on singing that song over and over.

While Wauk-jong-ka-ga was singing, the geese were dancing around counterclockwise, and whenever a goose would dance by Wauk-jong-ka-ga would grab it by the neck and wring its neck. Whenever he would wring somebody's neck he would put him into the

Marking Migration Patterns Of Birds On The Marsh

pot, and give a big warwhoop so the geese wouldn't know what was going on. And while he was putting the geese into the pot he just kept on singing the song: If you open your eyes, your eyes will be red. And of course no goose wanted to have red eyes so they just kept on dancing with their eyes shut.

But there was one female goose who thought . . . well, this is really very peculiar, very strange indeed. What can Wauk-jong-ka-ga be up to? So she thought she had better take just a little peek to see what was going on.

Well, she just couldn't see what reason the geese would have for making warwhoops like Wauk-jong-ka-ga said they were doing. So she opened her eyes just as she came close to Wauk-jong-ka-ga. And she saw what was going on.

He is wringing our necks, she cried. Wauk-jong-ka-ga is playing a trick on us and is killing us for his pot.

Then all the geese all flew right through the sides of the wigwam. And that is why the geese have red eyes today.

These geese reach their breeding grounds in late April, several weeks before the breakup of the major rivers. Their arrival so characterizes this month that it is known to the Cree Indians as *niskapesim* or goose moon. At this time there may still be several feet of snow in the bush.

While waiting for the snow and ice to clear from portions of the interior muskeg where they will nest, the early arriving geese fly back and forth between open spots along the rivers. To feed, they often resort to snow free areas in the muskeg where they consume sedges and berries remaining from the previous autumn. The Canadas are hardy birds and having put on a layer of fat before migration are easily able to sit out extended periods of severe weather. Even so, spring in the North is often capricious; late blizzards may force the advance flocks to retreat southward several times before they finally reach their northern destination. The great muskeg, as their breeding grounds have aptly been called, is a country nearly impassable to humans on foot. It is referred to by geographers as the Hudson Bay lowlands. Largely a waterlogged plain, 125,000 square miles in extent and lying only a few feet above sea level its surface varies from scattered blocks of stunted spruce and tamarack to large areas of bogs and pothole lakes. Because floating mats of sedges and grasses cover much of the water areas its appearance is often deceptive, and the Cree Indians who hunt and trap the muskeg may suddenly plunge hip deep in cold water.

However, not all of this lonely muskeg land of the North is attractive to pairs of nesting geese. Extensive bogs and large lakes in themselves usually do not constitute ideal nesting habitat; rather it is the patches of closely lying pothole lakes which have one or more small islands that appear to be the most attractive to nesting geese. In choosing such lakes, the mated pairs could not seek out more beautiful country for their summer sojourn.

Nest sites are frequently located on islands or islets, often close to woody vegetation and usually within a few feet of water. However, in some areas nests may be located out on waterlogged sedge — grass muskeg plains at considerable distance from any sizable pond or lake. Usually five to seven eggs are laid, with older birds producing larger clutches than birds nesting for the first time.

Although the factors which affect success of the nesting season have yet to be fully determined, weather conditions are undoubtedly very important. In some years the relative number of pairs that produce young may be as little as one-third that found in other years. However, because geese do not breed until two years of age, alternate annual variations in the age structure of a population are a normal phenomenon. For example, a bumper crop of young one year is certain to lower the percentage of geese of breeding age in the population the following year; conversely, two years later the addition of these geese to the breeding segment of the population results in another large crop of young which again reduces the percentage of geese of breeding age in the wintering population.

Canada geese are potentially long-lived, particularly the giant Canada goose for which there are numerous records of captive individuals living to forty and occasionally sixty years. However, in the wild, the average age due in part to heavy hunting is usually less than two or three years, although there are several instances of banded geese (B.C. interior) attaining at least twenty-two years.

Most of the Canadas that stop at Horicon weigh from seven to ten pounds, although the "giant" goose may weigh twenty. Only a few of the "giants" come to Horicon. The wingspread of a full-grown goose may be six feet.

Like most birds, the Canada goose is a curious mixture of sociability and intolerance of others of its kind. In breeding areas where the habitat is limited and nesting islands are scarce, as in some sections of the West, Canada geese will nest in close proximity. In the North, where lakes with islands are numerous, each pair will reserve a lake to itself, or in the case of the larger lakes, a bay or comparable section. Some may nest in boggy areas at a considerable distance from water, but most pairs are well separated from each other.

While the female undertakes the chore of incubating the eggs, the male stands guard somewhere in the vicinity. In the muskeg of northern Ontario, the male is usually seen several hundred yards from the nest. After a twenty-eight day period, during which time the female leaves the nest only briefly each day to feed, the eggs are hatched.

Soon after the young have hatched, the families seem to obey an urge to leave the nesting area. The adults are flightless at this time, of course, because they are moulting and growing new primary feathers on their wings. Those in the far interior of the muskeg, which constitute the bulk of the Mississippi Valley Flyway population, wander from lake to lake, feeding on grasses and sedges as they cross the intervening stretches of floating sedge mats. If the geese have nested near the sea coast, they often descend the rivers to more favorable coastal marsh and tundra feeding areas. When rapids are encountered, the birds travel overland to the next stretch of calm water.

A pair with their young of the year is an inseparable troupe, acting in unison almost as a single biological unit. In moving about, the female leads the way, followed by the young, with the gander bringing up the rear. When another goose family ventures too close and appears to be competing for the same feeding area, "battle formation" is assumed, the male acting as the head of a V-like phalanx, ready to do physical battle while the female and young assume threatening postures behind him. The gander literally defends the ground he and his family walk on, plus a few square feet of surrounding area. A fancied infringement of such indefinite moving territories by other geese may be the cause for a battle royal between the ganders of the two families. Curiously enough, the victor of such encounters can be predicted with such statistical assurance that if money were wagered, the observer "in the know" would be sure of a profit. In encounters between the ganders of two families, the psychology of strength in numbers seems to be the decisive factor, not the apparent size or weight of the antagonists. Thus male geese with large families almost always defeat males with small families, whether the ganders actually fight or merely threaten each other.

During this period of wandering, the young goslings grow their flight feathers, while the adults moult and regrow theirs. The family remains grounded until early August when the birds are ready to take to the air as a family unit. Some families remain inland, while others fly to the shores of Hudson and James Bay where they feed on berries and put on a layer of fat prior to their southward migration. There they are

joined by tens of thousands of blue and snow geese that have nested in the Arctic.

Some of the Canada geese linger on the shores of Hudson and James Bay until early October and then suddenly in the space of a few days they are gone. The inland geese tend to follow the north and south trending rivers. A swift flight, or one broken by stops, returns them to their autumn and winter quarters in the United States, which in most cases are Federal, State, and private refuges. Here, goose hunting is carefully regulated, with wardens employed to see that laws and regulations are observed. The geese are strong eaters, when food is available and consume about a half-pound per day per bird . . . a lot of corn when a half-million hungry geese are eyeing cornfields.

The State of Wisconsin has tried different studies to see whether Canada geese could be managed so they nested at Horicon. The State men would like to produce about 10,000 "homegrown" Canada geese on State refuges, eventually. Meanwhile the migration to Horicon gets larger and larger. The managers would like to get the geese trained to go to other refuges too. So far they like Horicon best.

I went one morning up to the main State Head-quarters office to see James Bell, director of the wild-life area. I found him extremely enthusiastic about his work, and about the whole atmosphere of the Marsh. Once I got him started, he just kept going because there was so much to tell. Bell, a graduate of the University of Michigan, is still a young man, though he's been at the Marsh since 1952. He wore the morning I saw him a red plaid shirt, Khaki trous-ers, work boots. And he sat back very relaxed at his desk. Behind him, on the wall, was a dart board with a single dart, which, no doubt, he used to decide the day's activities. The one target area said: What shall I do first?

Bell explained that he loved the natural state of the Marsh and found it constantly beautiful: It's been changed somewhat by man, he said, you know about the drainage ditches, and the dam at Horicon that floods it; and you have the tree and shrub plantings all put in by man, but they blend in well with the natural setting. Since a hundred years ago, if you want to go that far back, there have been radical changes in the Marsh. Originally, and just to review it again, this was a Marsh as a result of the glacial ac-tion. Then man built the dam for a gristmill, pretty close to the site of the present dam, and as a result it went back to lake again. Then drainage interests in the early 1900s decided that the Marsh could be re-claimed for agriculture. So they drained it and after they did, you could go out here . . . the old-timers used to go out to this point here to cut Marsh hay. You've talked with Pete Feucht about that. Then, to get better hay they would burn the Marsh over peri-odically, and the fire would burn down into the peat; sometimes you couldn't stand the smoke in Mayville or Horicon. Then it was reclaimed after the drainage project proved a failure, and the state and federal governments purchased the land, and the dam was built about 1930 in Horicon to flood it.

'Course it has silted in a lot over the years because of erosion from the watershed.

The patterns of wildlife have changed, too, espe-cially as concerns the Canada goose. In the 1940s, for example, only a few hundred geese flew over, much less stopped. We know back in the early days, the late 1800s and early 1900s, geese used to breed in here. Disturbances drove them away, and they went up Hudson and James Bay area to breed.

After the Federal Government purchased the Ref-uge at the north end, they began planting food — corn, rye, buckwheat. The first thing they did was to close the area to hunting. The geese had sanctuary.

And that helped, and they began to stop here in ever-increasing numbers. In 1950 they built the dike across the south end of the Refuge, then the geese had all three requirements. We are at the point now where the geese have become a terrific problem; we'd be much happier here with 50,000 geese instead of the 250,000 we had last fall. The Federal people figure that food for 50,000 geese is the limit they can plant in the Refuge. The geese now consume the food in the Refuge very quickly, then they go out and feed in the farmers' fields, and this makes the farmer mad. Prior to this huge concentration of geese and with uncontrolled hunting pressure the geese were more or less kept under control insofar as crop depredations are concerned. Farmers on the periphery of the Marsh had quite an income renting their land to hunters or leasing blinds by the season. The price kept going up, up, and up as the goose hunting got better.

In 1959 or '60 the Federal Government felt that the Marsh was being overshot and they assigned a quota, not only to Wisconsin but to Illinois. The last two years the quota has been 35,000 geese for the whole state of Wisconsin so that only so many birds can be taken and the number of hunters is limited, and the farmers aren't getting that revenue that they formerly had. They see all these geese here — more than ever — they can't shoot anymore, and they find it hard to understand, because the birds leave here and go to Illinois and are taken down there; they get right unhappy sometimes.

I asked Bell whether the Horicon Marsh was different from other marsh areas in the country.

Well, he said, it all depends on the type of refuge you're talking about. If you're talking about a waterfowl refuge in this part of the country, Horicon would be pretty much typical. Depends, too, on the kind of habitat you're dealing with. Down in the southern states, for example, I expect the refuges are more grassland. The cover here is very similar to what you'd see at the Eldorado Marsh area at Fond du Lac.

What do you do this time of year? I asked. What is your seasonal work?

Well, this winter, with the amount of snow that we've had, we've been very busy distributing corn to the cooperators. When I say cooperators, I mean farmers, individuals, and members of sportsmen's clubs, feeding the pheasants. Last fall the conditions of harvest were very good, and the bulk of the corn was picked, and some of the fields were plowed.

What we do, we haul corn to a central delivery point, and it is parceled out to the various cooperators from there. For example, there is a sportsmen's club down at Hustisford; they come up here and get the corn and take it down there, and the members distribute it to the birds.

Got a case over at Beaver Dam where the conservation club is feeding, and then you got a sport shop there; we take corn to them and they find an outlet for it. We've got about eighteen or twenty distribution points in Dodge, Jefferson and Columbia counties, our area.

Winter of 1971 we fed forty-four tons of corn and it was only February. This has kept our field crew very busy. We get the bulk of the corn from our wildlife areas. We leave some of it standing for the pheasants and the deer, and the rest goes into storage here, or if we have any surplus we send it down to the game farm.

The farming aspects, then, are under your direction here, too? I said.

In the wildlife area we have what we call a share-cropping program. Local farmers do the actual planting and harvesting, and for their efforts with corn grown on Marsh land they get two-thirds of the crop. We take the other third.

What happens in the spring on the Marsh? I asked.

Well, in the spring, we will be planting trees and shrubs. We do extension work with landowners who want to improve their property for wildlife, cover, this kind of thing. The spring is devoted to improvement, and the summer is about the same, posting our wildlife areas, construction of dikes. Those have to be maintained.

I inquired about pesticides. We never use pesticides, Bell said. The fall is the busy season. We have to see that the areas are fully posted, make the closed area plain, and there are areas we flood in the fall of the year for ducks. Throughout the fall, too, we are busy with patrolling, checking into conditions. Our men have warden credentials; they watch for game law violations. And of course we are very busy, have been the past several years, with goose damage. We have a law here in Wisconsin that does pay for crop damage. The first thing we do if we get a complaint is go out there with exploding devices to help the farmer try to get rid of the geese. But if we have a wet fall as we had in '65 and '67 then the geese have already been in there and have caused damage before we can get there. Then we have to investigate the amount of damage and come to an agreement with him (the farmer), and make out claim forms.

Are there instances where geese have actually cleaned a farmer out? I asked.

Yes, we've had situations like that. I think the highest claim we ever had was in 1965, the first year they passed this crop damage law. The damage amounted to some $9,000. Was a large farm called the Flying Dollar Ranch. Recent years we have kept the claims down to an average of about $70 or $80 or a few hundred.

How do the State men get along with the Federal men at the other end of the Marsh? Any problems?

We don't see eye to eye on all aspects of goose management, Bell said, thoughtfully, but we keep it at least on a friendly basis. We aren't at each other's throats.

What is their point of view that differs from yours?

Well, they would like (and so would we) to see fewer geese here on the Marsh. What we can't agree on is the method of reducing the number. We would like to see, for example, the possibility of opening the Refuge to hunting. And you probably read about the "hazing" that was done in '66. That wasn't successful. We brought up the hunting possibility at one of our recent meetings. But they were afraid that the geese, driven off the Marsh, would only light on the farmers' fields, cause crop damage, and antagonize the landowners. Well, that would be true, except that we would put a concentration of hunters not only in the Refuge but also on private land, and maybe with a week of hunting pressure we could drive some of the birds out of here. There was mention of not planting any food in the Refuge, except along Highway 49 east of Waupun, where the sightseers could see the birds. One of the things the Federal men propose is draining the Marsh, and having very little water in the area. But that would mean that we would have to drain our portion of the Marsh, so there really is no concrete suggestion that assures results. No one's in a position to say, "I know this will work."

Wings

They think, too, that the State ought to acquire some more areas outside the Marsh where we could siphon off some of these birds. Well, we've been doing that. We've got Eldorado Marsh in Fond du Lac County, and Grand River Marsh at Kingston in Green Lake is well along. The dam has been built there, and this will attract a lot of geese. This isn't gonna be right now. It will take several years, and there are also refuges in Calumet, Manitowoc, Juneau, Dodge, Shawano, Columbia, Wood, Sheboygan, and Washington Counties that will help.

No matter how many years it will take, the Horicon Marsh will remain a fascinating place. Waupun, at the north edge, is already making great commercial good of the Marsh and the wild geese by having Wild Goose Festival Days in October. The city of Horicon also has a Horicon Marsh Days Festival in July. The citizens have organized a fine arts council at Waupun, and combine commercial enterprise with art shows and a very unique Middle West sculpture show that attracts more than 300 major sculptors.

Sightseers viewing the birds can go to Waupun and see the famous End of the Trail statue, which conveys the spirit of the defeat of the red man and was the most famous work of the great American sculptor James Earle Fraser (1876-1953). Clarence Addison Shaler, Waupun inventor and manufacturer, who made most of his money through the invention of the cold patch for automobile inner tubes, had a major interest in sculpture. He saw Fraser's End of the Trail at the 1914 Exposition in San Francisco. There the work was in plaster, and Shaler asked Fraser whether he could provide a bronze of the famous figure of the dejected horse and defeated Indian for a setting in Waupun. Fraser replied that he could do so. The formal unveiling took place on June 23, 1929.

172

Law enforcement is a very important part of the management program at the Marsh. Pat Burhans has been head warden at the State of Wisconsin Marsh Headquarters for about fifteen years. He was formerly located up in northern Wisconsin and moved to Horicon in 1957. Pat didn't know too much about the Great Marsh when he came down from the north. He had a choice of coming to Green Bay or Horicon and he picked Horicon. I guess he thought the name sounded more romantic.

When Pat arrived at the Marsh in 1957, the geese were just beginning to come in large numbers. There were probably about 50,000 geese that were regularly visiting the Marsh at that time, and the wardens managed an organized goose hunt. They had two check stations, one on each side of the Refuge, and they had blinds around the perimeter of the Marsh. Every morning they had people lined up and waiting for the hunt to get under way. Each hunter could bring two guests. They allowed three men in a blind and they each paid a dollar. When a party of hunters would get their goose quota, they would have to leave the blind, and wardens would refill the blinds from those who were standing in line. The blinds were filled first from those hunters who had made blind reservations, but there would be several hundred cars, filled with hopeful hunters who wanted to occupy a blind when it was vacated.

The bag limit was two at that time. No duck shooting was allowed. Usually the hunters were well behaved, but once in a while some fellow might be drinking, or just feel mean, and he wouldn't leave, or sometimes a hunter would shoot more than two geese. Then the wardens would have to come and deal with the situation. Everybody wanted to kill a Canada goose, and sometimes Pat, standing and watching the lines of hunters, would wonder what it was that impelled men to want to hunt the birds when it was so easy to kill them. Sometimes the geese would come over so thick that about all a hunter would have to do was to stick his gun up in the air and fire. Not much skill was required.

The rules got more stringent as the flock built. Hunters in a blind were soon limited to two, and then one goose was all that could be taken by a hunter. The wardens changed from trying consciously to increase the goose harvest, to trying to control it. And finally they abandoned the managed hunt in the mid '60s, and went to the present system where they have all hunting done on the privately owned areas around the perimeter of the Marsh.

Perhaps the stimulus for the stricter rules was given

by the Federal Government when a quota was imposed . . . only a certain total number of wild geese could be killed; and then the State went to a voluntary registration system. This system depended on the hunter stopping at a roadside box or a tavern, or a filling station, to pick up a registration tag. He filled out a card indicating what his kill had been. The wardens soon knew that they weren't getting very good compliance with this voluntary system . . . only about sixty percent of the hunters actually stopped to get a card.

The goose flock was increasing so rapidly that the State and Federal men got worried. There were simply more geese coming than could be fed, and they were spilling over, off of the Refuge, onto the farmlands around the Marsh. Irate farmers complained that their crops were being destroyed, that they would be ruined by the ravenous birds that could reach upward a good four feet and strip off every ear of corn on a stalk. At the same time, there could be no uncontrolled slaughter of the birds, though farmers often threatened to do just that.

Finally, the Federal men at the north end of the Marsh proposed a great program of hazing the birds away . . . simply scaring them so that they would

Search For Lost Hunters On Horicon Marsh

away . . . simply scaring them so that they would never again stop at the Horicon Marsh . . . something had to be tried, anyway, because now there were 100,000 geese stopping at Horicon. That program, which is elsewhere described, of hazing the geese, was a fiasco. The Federal men tried to scare away the birds with airplanes and cannon planted in the fields, but the State of Wisconsin soon issued injunctions and the State served the injunctions.

The whole thing backfired in a grand and terrible way. For about two weeks the Federal men had been hazing the birds. As a result the birds flew out of the Marsh area during the day, or while the hazing was going on. And then they flew back into the Marsh at night to roost . . . as though they actually knew that the airplanes which were scaring them would not fly at night.

But then the goose season opened, and hunters around the perimeter of the Marsh had the lushest hunting ever experienced. The hazed birds simply flew out over the guns of the hunters, and in two and a half days approximately 34,000 geese were slaughtered. The goose season came to abrupt end; it had lasted just two and a half days. It was a great free-for-all. Pat says he never saw anything like it. It must have been the greatest goose shoot in history. Anybody who felt like killing a goose would simply go and do it. Pat went out in his car with another warden one day to check some hunters on the opening day of this sea-

son. This warden Pat was with, was a stranger. He had just been sent to Horicon to help the State men. As this warden and Pat were walking through an area near the Marsh, the visiting warden disappeared over a hill, and a few minutes later he came back carrying a goose. Pat, of course, asked what happened. Why was the warden carrying a goose? Did he confiscate it from a hunter?

Well, the warden said, I walked up to a guy to check his license, and the feller shoved his shotgun into my hands and says, here, shoot yourself a goose. And so I done just that!

Pat just stood and looked at the clouds of geese flying overhead and couldn't think of a thing to say, except, Well, I'll be damned.

Hunters took away uncounted scores of geese. They would get one and run to the car with it, and then come back and shoot another. The wardens made many arrests, but it hardly helped the total situation at all. There were just too many hunters and too many birds and far, far too much confusion.

Nowadays the rules are plainer, and hunters are easier to control. Hunting has been prohibited from any roads and from any railroad tracks, in the entire Horicon Marsh zone. This eliminates the big lineup of sportsmen waiting to bag a goose on the roads. Then there is a smaller control zone called the intensive management zone where supervision is closer; and wardens have now gone to blind spacing . . . that

174

means that every landowner around the Marsh can have one blind for every twenty acres of land. The blinds have to be two hundred yards apart, a hundred yards from a neighboring property, seventy-five yards back away from the Federal refuge if the blinds are on that end of the Marsh. Not more than two people can occupy a blind at any given time. This blind spacing has been a good enforcement tool for Pat and his wardens, because, over the years, they have become familiar with the locations of all the blinds on a property. The wardens are able to pinpoint where a hunter is located, and they can keep a much sharper eye on him. If the wardens hear shooting or see birds dropping in an area where they know there are no blinds, they move in fast and apprehend the lawbreakers.

The farmers cooperate very well as a rule. There are some farmers, though, who are bitter, mostly because they feel that the laws are too strict. Their reasoning is that there are so many birds, and why impose such strict laws? Wouldn't it be much better to kill a lot more geese, and get rid of some of the flock . . . Also, the farmers, too, are limited to the killing of just a single goose, even if it is on his own land. Many farmers cannot see the justice of this law, especially when the birds feed at will upon his corn crop.

Some farmers living around the Marsh, however, have a very good thing financially. The State men checked with the Federal Bureau of Internal Revenue one year. They investigated the returns of one farmer who had a number of blinds on his farm. This man had reported an income of more than $60,000, just from his goose blinds over a ten-year period.

When the geese first got fairly numerous, most of the farmers leased their land. The hunter could kill an unlimited number of birds. He was limited to a daily bag; but if there was a forty-day season he could hunt every day. Consequently most of the choice hunting spots were leased up by big companies, to entertain friends and business associates. Hunting for the ordinary citizen around the Marsh, or the man on the street, was practically unknown. The local hunters from Horicon and Burnett and Mayville didn't get much chance to kill a goose.

When the State went to the tagging system where the hunter could only kill one goose a year, immediately it became no longer lucrative for the farmers to lease their farms. Why would they want to lease a whole farm when only one goose might be killed? Consequently, it forced most of the farmers who were interested in having hunters around, to go to renting blinds by the day. They had to begin to advertise, and this procedure made the wild goose available to the man in the street. The local guy, or the guy from Milwaukee, if he was fortunate enough to get a tag and a permit, could drive out to the Marsh area and find a place where he could rent a blind, for somewhere between five and fifteen dollars, and be reasonably assured of killing a bird.

That's about what the situation is today. There are movements to change this system. Some people would like to dispose of the goose tags and go to what they call "on the farm" registration. This is what they do in Illinois. The people who rent the blinds keep track of the kill. The season would close, under that system, when a fixed number of birds have been taken.

In Illinois, though, they only deal with sixty or seventy operations and the whole problem of goose hunting is much smaller. In Wisconsin the authorities deal with some 900 landowners and other individuals, and it would be very cumbersome to go to any kind of "on the farm" system.

Around the Marsh, on many of the farms, if the hunter isn't successful in getting a goose, the farmer will allow him to come back and try again with no charge.

The farmers, of course, aren't able to make nearly as much by renting out blinds by the day, as when they were able to lease the whole farm. Pat knew a farmer who, in the old farm-lease days, said to the representative of a large Milwaukee business concern, "Well, sir, you know, I been thinking of gettin' one of them $3500 John Deere tractors, and if somebody would just happen to drive up here with one . . . well, that feller could have the farm."

And as a result the farmers, some of them only, are bitter because they lost some of this revenue. But there are a lot of other farmers who aren't the least bit interested in hunting. They want to farm. They don't want any geese. They don't want any programs that involve them. They just want to be left alone. Especially they don't want 20,000 Canada geese walking around feeding in their best cornfield. In fact, some farmers actually hate the geese!

And some of the farmers like to hunt, too, and there's no provision for the individual farmers to do any more hunting than anyone else. Consequently the farmers get one goose tag, and this makes them unhappy.

The farmers aren't the group, though, that gives the wardens their most difficulty. With the random selection such as the State employed with the goose tag, they issued 23,000 tags in 1971, and there were 49,000 applicants! The Marsh is within an hour's drive of two million people in Milwaukee and the Fox River Valley complex; and the whole Marsh area does have a large influx. These urban visitors sometimes cause trouble. They don't want, often, to put the tags on the geese. They just want to get the goose out so they can reuse the tag. Another problem that plagues the wardens are the hunters who come in and who don't know where to go to rent a blind. Consequently, they see a flock of geese flying over, and they just get out

of the car and bang away.

But the wardens have trouble with the local poachers, too. They're more familiar with the local situation, and the type of violation committed by the local folks is more deliberate. They really plan to break the law . . . it's not just a spur of the minute thing with them.

In 1970, in the fall, the wardens made 243 arrests. In 1971 it was higher; and the number of arrests is going up each year. The more vulnerable the game gets the more the occasional unscrupulous hunter will take advantage. Also, the number of arrests varies with the amount of money budgeted by the Marsh headquarters to hire seasonal wardens. The more wardens, the more arrests.

Pat divides the Marsh area up into sectors and assigns his men to each sector. They all go on patrol, check licenses. And they do rely on informers. They quite often get calls from people who usually don't say who they are, and the informers will tell Pat where an overkill of geese is going to be attempted. Pat operates the whole Marsh enforcement from his 176

Citizens Tree Planting On The Marsh

office in the headquarters building. He has radio contact with all his men at all times. The lower end of the Marsh is public hunting grounds. In that area it is free to anyone. And most of the hunting in the public grounds is done from boats. Pat keeps two wardens doing nothing but checking on the boats.

After arrests are made the paper work gets intense. Pat has to have incident reports on each individual case, and Doug Radke, who is the Dodge County warden, is the court officer. He goes into court each Wednesday morning and handles all the cases. A lot of the cases are bond forfeitures . . . they have the choice of forfeiting the bond, or having their case heard in court. The evidence is one of the biggest problems. In order to preserve the game that has been illegally shot, there is a large walk-in freezer in the basement of the headquarters. The sorry part is that after the cases are all disposed of, there is very little that can be done with the killed birds . . . about all the State men can do is bury them. Last fall almost 200 Canada geese were quietly buried. The evidence has to be available in case there is a trial, and a not-guilty plea. The shot birds have to be preserved carefully, because as Pat says, the State doesn't always win the cases. In case the arrested party wins, then the State has to return the bird to the hunter . . . ostensibly in good condition.

By Federal law no one is allowed to sell waterfowl.

The state can only give confiscated birds to public institutions: hospitals, schools and prisons . . . and it's been almost impossible to give the game away. The birds are not inspected; the institutions don't have the facilities to prepare and handle the game; there is always a chance that a bird might be diseased . . . and when the evidence is finished the wardens turn the game first over to the research people who find out what they can of information that might help the program of wildlife preservation.

Pat does encounter some characters in the law enforcement business. With some of the violaters it gets to be a regular game. They had one old guy from over by Burnett who was getting to the place where the officers didn't even want to arrest him any more. He was in his late seventies, but he was still pretty spry. The wardens, in order to sort of protect the old man, would drop by his place two or three times a day just to see what he was up to. Of course the old fellow would get pretty riled up. He figured the officers had no right to check up on his comings and goings . . . which would have been right except that he was almost always on his way to do some illegal hunting.

Pat was going by his place one evening just as it was getting dark. There was a big tree out in the yard where the old man always hung up his buck if he got one. There wasn't any deer hanging in the tree; Pat

Fish Planting

could see that, so he went his way. But real early the next morning Pat was going by the place again, and this time there was a big deer hanging in the tree. Obviously the deer had been killed in the night. Pat stopped, got out of the car and walked up into the yard and stood there looking at the deer. The old guy came out and he had his shotgun and it was pointed at Pat. The man said, Warden, you touch that buck and I'll drop you where you stand!

Where'd you get the buck, Pat says.

You're gonna say I shot it last night in the Refuge, ain't you?

You got any proof you didn't? Pat says.

You got any proof I did? says the man.

Nope, Pat says, and gets in his car and drives away.

One day Pat watched a suspicious-acting fellow through his telescope. The hunter shot a goose, and then, looking all around to be sure he wasn't being watched, he went over to a fence row and skinned the bird, put the meat in a plastic bag, and put it in his shirt. Then he looked around some more and finally

hunter paid a nice fine.

Pat had one hunter that shot a goose and put it in a five-gallon pail and covered it with sand. He says, when Pat asked him what he had in the can, Why warden, I just got a can of sand in here. For my little kid, he says, who likes to play in the sand.

So does the judge, Pat says.

They've had hunters put illegal birds in bags of decoys, and they've tried about every dodge there is to try. Pat knows every wrinkle.

The State has a nice cabin out in the Marsh, away in, that is quite livable. And Pat will sometimes keep men living out there full-time, so they can watch for violators. Once he had a crew living in the Marsh for seventy days straight. The men would come into town a couple of times a week with the evidence, and the confiscated guns. The hunters' guns will always be returned to the hunter unless the fellow has been in trouble a lot before, or unless there is something very unusual about the violation. The court has the right to confiscate the guns if it sees fit, and in that case

Bird Marked For Migration Study Taking Off

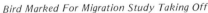

Bird Marked For Migration Study Taking Off

walked out to the road.

Pat strolled down to meet him, and says, Had any luck?

Nope.

They gossiped a while and suddenly Pat poked the fellow in the stomach and says, What you got in there?

Why I got a goose in here, the hunter says. And if you got an oven, warden, you and me'll have the damndest goose feed you ever seen. I know how to cook it, he says.

Well, they didn't have any goose dinner, and the

the guns are taken to Madison and sold.

Pat says that a good warden can never be bitter about his work. He has to be able to deal with the people. In other words, a warden shouldn't just come running up to a violating hunter and yell, All right, you son-of-a-bitch, you're going to jail! That's not the best way to do it. You have to treat the hunters with dignity and try to impress on them that they really have done something wrong, and that they are going to have to pay a penalty.

Pat and his family really like to live around the Great Marsh. They never get tired of looking at the geese, or of going out on the Marsh, and of showing the Marsh to somebody else. Pat has developed a very strong feeling of his personal relationship to the Marsh. It is a part of him, now. He's gotten really interested in the history of the Marsh, too, and spends quite a bit of his spare time collecting mementoes of the older days . . . historic decoys, and old maps. The Marsh has been growing on Pat ever since he moved to be a part of it.

In the early days when Pat first came, pollution was a very serious threat. A lot of industrial waste was coming in big quantities from metal manufacturing companies in Waupun . . . cyanide and other chemicals . . . and to the east branch of the Rock River was coming chemical pollution from factories at Mayville and Allenton. There has been great accomplishment in cleaning up this pollution. The companies have cooperated, and the mismanagement of the watershed is probably the big problem now . . . the cow and the plow. Pat really thinks the hunter isn't the chief menace to the Marsh and the wildlife... it's more likely to be the local pollutors who might do the final damage. Actually they have the hunter problem very well controlled.

Pat believes, though, that there really are too many geese stopping at the Horicon Marsh. The total number of geese in the flyways is too great. Fifty thousand geese would probably be the right number for Horicon.

This year the headquarters received about ninety complaints from farmers who complained of crop depletion. They were eating too much farm corn. Maybe ninety doesn't seem like so many complaints, but if there should happen to be a real wet fall, when the farmers couldn't get into the fields to harvest the corn at the peak periods, in October and early November, there could be mammoth losses to the farmers; and the farmer never has the right to just go out and slaughter the birds. There have been some incidents where a farmer, angry beyond control, has gone out and killed some geese in his cornfield. Then usually he has simply called the Marsh Headquarters and said: Come out and get 'em. They're here.

The Bureau of Game Management have intensified their efforts to try to help the farmers with the goose-corn problem. They have equipment for scaring the geese, and they lend this to the farmers, or more often, go out and put the equipment into the fields themselves.

They have, mostly, a sort of cannon. I went with Pat out to the shop to see the rows of these noise-makers. On a high shelf stood perhaps fifty devices that looked much like a small, old-fashioned cannon. At the breech was a chamber which was piped to a propane gas drum. The gas collects in these chambers and a revolving hammer falls, at the proper moment, against a flint and sparks set off the gas. The cannon makes an exceedingly loud, but harmless, noise. It scares hell out of any goose within a couple of hundred yards . . . especially if the birds aren't used to the noise. When they do get used to the cannon, Pat says, he has seen them feeding only a few yards away, and they don't fly even when the explosion takes place. They just go right on stripping down the stalks of farmer's corn.

Between the State and the Federal Headquarters there are about a hundred of these cannon on the Marsh. Usually a farmer will call and say that he wants a cannon or so installed in his fields. The Game Managers go out and set the cannon up. One cannon is good for about each twenty acres. The farmer has to keep the cannons operating. The State will pay damages to farmers when it is established that damage is caused by grazing waterfowl. The Federal government, however, will not pay any damages. The Federal men claim that if they opened up to claims for damages at Horicon, every wildlife area in the whole country would be deluged with claims of nearby residents.

The wild-goose-scaring project is really a large one. I was amazed at the time the wardens give to just frightening geese. The men circulate during the fall, all around the area, and each warden has a twelve-gauge shotgun, and a supply of special shells — loaded with a charge just big enough to propel a small bomb from the gun. Up about a hundred feet or so the bomb explodes, and the geese are supposed to rise in terror from the field, or if flying over, they are supposed to get the hell out of the area. Usually they don't . . . or at least they don't go very far. Part of the problem is that the farmers who ought to be doing most of the frightening-away work, just aren't willing to spend much time. The farmers just don't want to lie out in the cornfields all day long. So the geese get in their licks too.

It is interesting to note the number of sportsmen's groups related to the Marsh. These are:

Horicon Marsh Bowmen, Inc.
Kenneth Metzdorf, Pres.

Theresa Rod & Gun Club, Inc.
Reuben F. Schuster
R 1, Allen Road
Theresa 53091

Atwater Conservation Club, Inc.
John Messer
R 1
Burnett 53922

Brownsville Sportsmen's Club
Donald Collien
Box 95
Brownsville 53006

Clyman Sportsmen's Club
Harold Lehman
Clyman 53016

Field & Stream Sportsmen's Club
Kenneth C. Dittberner
Box 53
Knowles 53043

Hustisford Rod & Gun Club
Glen Wiesmueller
R 1, Box 269
Neosho 53059

Iron Ridge Sports & Athletic Club
Allen Schmoldt
Box 139
Iron Ridge 53035

Izaak Walton-Watertown
Walton Otto
1422 Willow St.
Watertown 53094

Sunflower Banquet

PROHIBITION DAYS AT THE MARSH

When I was first hired by the Department, said Ed Lechner, I had to be up in the old barn. It was in the winter time. We had those days a fellow named Pokey Zimmerman. He was quite a sportsman, and he came up to the headquarters to get some corn for the deer. He lived out by Beaver Dam Lake. And he asked my boss, Jim Bell, and they came out to the shop, and Jim says, Ed, scoop up a couple of sacks of corn for Pokey. And Pokey looks at me and says, Why Jim, what are you doin' with Ed here? Jim says, Why Ed works here. And Pokey says, What'd you want with him for? He's one of the biggest outlaws in the Horicon Marsh! Well, Jim just says, well, Pokey, if you can't catch 'em, you got to hire 'em! And that's the way I started to work at the State Wildlife Refuge. But old Pokey forgot one thing. After I grew up I started to straighten up, as the law started comin' in. It was just a part of the whole change. Like the change from Model Ts to Model As . . . when the Model A come in we started gettin' people from away off out there in the Marsh area to hunt and fish. Before then it was just folks from locally . . . Kekoskee and Mayville. In the older days they came by train, and stayed a week or so at the hunting lodges . . . the Diana and the Strooks. And those days all the local people fished and hunted just when they wanted to. We fished with spears and we fished with nets. And we went hunting with ferrets . . . I kept a couple of ferrets at the mill I had, to catch the rats, and I sometimes used them for hunting rabbits. Sure.

And then old Bill Reeby came. He was one of the wardens. I think there had been two wardens before old Bill . . . they was Dietrich and old George Hall. But anyway old Bill would come in the mill and he'd say, Well boys, you're goin' to have to cut out using fish nets. You're going to have to start fishing with a hook and line. And I had some nets hanging there in the mill and old Bill took his jackknife and cut down the nets, and he cut them up some, too. And well, old Bill hadn't hardly gone back to Horicon before we had the nets up and patched together again. Next time he came . . . and this is just to give you an idea, how this thing of the law just kept closing in on you . . . Next time old Bill came he took the nets outside and put kerosene on 'em and burned 'em. Well, we couldn't afford to buy new nets. We was all poor them days. So we began to be a little more careful . . . a little of this and a little of that, you know, and pretty soon, before we knew we was doing it, we got to obeying the law. There got to be a season, and before that we never knew there was any huntin' season. We just made our own season. It was just the way of life around the Marsh, that's what it was, and I guess none of us thought of ourselves as outlaws, either.

We were driving Ed's State truck as we went from place to place to see this farmer or that who might interpret the Marsh. About noon it started to sleet, and the sleet changed afterwhile to a steady rain that froze as it hit, and soon the truck was covered with an armour of ice, and the road was glistening and hard with a thick coat of just about completely dangerous slipperiness.

Gard: The rain sure is comin' down. Guess it'll take me a while to drive back to Madison. I'll leave a little earlier than I have planned to.

(As we drove, and Ed took more and more care and the road got worse, and the windshield ice covered so thick it almost stopped the wipers, so that Ed had to get out and clean them. As he worked, I got to thinking that maybe there had been some angles or aspects of life around the Marsh that a person wouldn't often tell about.)

Gard: Now, Ed, I'm going to ask you a few questions about what you might call some of the seamier sides of life around the Horicon Marsh. (Ed looked around at me, taking his eyes off the slippery road for a second. His eyes were both wary and still, quick with interest.) OK, Ed said, Shoot.

Gard: Wherever you get a collection of farming people and outdoor people and sportsmen, especially, perhaps sportsmen, I assume they are apt to be a little hot-blooded or red-blooded, or just lusty, or whatever you want to call it. So I assume there must have been some prostitutes around this part of the country, just as there were and perhaps are, around most every other part. You know the stories about Hurley — true or false. I guess, in a word, Ed, there must have been some houses where the country boys could go if they wanted to find a woman companion . . . were there any such?

Lechner: Well, years ago, when I was a young man, I would have to say yes. I couldn't judge now, though.

Gard: I was speaking of the past.

Lechner: Minnesota Junction.

Gard: Where is Minnesota Junction?

Lechner: Oh, just west of Horicon . . . about three, four miles. Northwestern gets crossed by the Chicago, St. Paul and Milwaukee, there.

Gard: What was Minnesota Junction like, and why did they call it Minnesota Junction?

Lechner: There was a little railroad station there. The trains used to go through there, lot of 'em. Both tracks. I guess it might have been just because the guys around here thought of those Swedes up in Minnesota as being the . . . well, the kind of guy who would go to a place like they had in Minnesota Junc-

Marsh Cobwebs Also Spin Memories

Marsh Winds Create Their Own Designs

186

tion. If that answers your question.

There were three places there in those old days. The houses were kind of nondescript and maybe unmentionable. They weren't palace houses or anything like that. Just pretty simple places, but I suppose they seemed like really somethin' special to a few of the local boys because the girls dressed up pretty fancy with bows and ribbons and little gewgaws and stuff like that, you know. The biggest popularity of those places at Minnesota Junction, though, came during Prohibition. In a very small way, the vice that infected the great cities crept upstate, and polluted our rural neighborhoods, too. I know of several Chicago gangsters who had places up in here not too far from the Marsh, especially when they was makin' alcohol to ship down to the city.

'Course, some of our fellows went over to Fond du Lac which was pretty fairly wide open, and there was a place at Addison on Highway 175 . . . was three buildings there, a bar, and . . . you know what I mean, some women, and used to be called Addison's.

Over at Minnesota Junction, a woman run one place, Charlie run the other place, and Til run the other place. This woman had quite a joint. She was a real boss lady and she had five or six girls there. All of these people came in from somewhere else. None of 'em were natives. I suppose she come up from Chicago. At the beginning it was all right, but then, like the tavern business in Prohibition, it tightened up. That's about all I know of the houses, but a whole lot of folks made beer.

When I was a young man my dad used to farm me out to teach these local farmers how to make home brew. I would go to a farm and stay for a day or two and show 'em how to cook it and ferment it. Dad was a real generous guy and wouldn't take a thing, not let me take anything for teaching these farmers. Dad was always helping somebody, and I guess he thought that it was a real help to a fellow to teach him how to make decent home brew . . . not this awful rot-gut stuff. Well, he had that skill, and that was what he could offer. So why not?

Gard: He taught all the recipes he knew?

Lechner: That he did.

Gard: I guess you could make a pretty fine home brew even now, if you wanted to.

Lechner: That I could. That I could. You see, during Prohibition, when this thing really started going, and we had it set up, we had a wort brewery. We called it wort. That was over at Waupun. You were allowed to make wort. That wasn't against the law. They put it in five-gallon cans.

Gard: How do you spell it?

Lechner: W-o-r-t. This we would put right into our vats. We would set it with brewers yeast. It would

start to work. At first we were able to get syrup . . . in cans, you know. You would pour this syrup into a tub filled with water and you would boil this with hops and then you'd pour this stuff in your jar, and add some water to it and let it cool down, then add some yeast . . . this was the way you made your home brew. But this wort stuff, it by-passed all of that. You just bought the wort at the brewery, and it saved a lot of time for the process. A fellow could make up a batch of home brew in no time, using wort. It was a lucrative business. Feller name of Arnold Peterson, he run this brewery over at Waupun, and he had a great big truck, a semi, hauling this stuff, he had so much demand for it. In fact one of the old truck drivers is still alive over to Waupun. He used to haul this wort to our brewery. All we had to do was set the yeast. Our beer vats held twelve-and-a-half barrels to a vat. We had those vats down in the basement in our little wildcat brewery. Well, this Arnold Peterson, this guy who made the wort, he got to be a wealthy man. He was working within the law. Wasn't a thing the law could do to him. Wort wasn't intoxicating. Didn't contain any alcohol.

I can tell you a story, too. This Arnold Peterson was quite an operator. He was a multimillionaire during World War I and he lost all that money in speculation. Then he started this wort business in Waupun. He made another big fortune in that. But then beer and whiskey came back and he had invested all his money in these little breweries . . . the whole country was peppered with 'em. And of course when Prohibition went out, and they began to get the competition of all the big fellows: Pabst, Schlitz, Budweiser, and so forth, well Arnold Peterson just went broke. All the little breweries that had been so important all through the early periods of the settlement and early development of the country, had to shut down. The shells of a lot of them are still there, in the small places, to remind those who know or care, how it was in the old days, when life was quieter, more leisurely, and a man could jog along in a horse-drawn buggy, or in a lumber wagon, and if the day was hot he could stop most anyplace and have a beer or two brewed right there at home. The Germans and the Bohemians, they made most of the beer. They was great brewmasters. Still are I suppose.

Well anyhow, I was running this wildcat brewery over to Mayville. My dad was dead, and Arnold Peterson called up one day and says, I want you over at Waupun. He says, Drop everything and come over. I drives over there, and just to give you an idea what kind of operator he was, this will give you an idea how political things were then.

I walked into his office. He always chewed a flat cigar. Never smoked it. Just chewed it. He says sit down. Poured me a little drink. He says to some other men who was in there too, he says, I want you fellows to take a look at the kid. He is supporting his mother over at Mayville. I wanted you to see what he looks like. He says to the other men who were kind of slouching against the wall . . . had their hats on. One of 'em had a hand in a side coat pocket. Made me kind of nervous. They was watching me so close and all.

Arnold says, I want you to lay off of the kid's brewery. And right then I got the situation. Was a fix that Arnold had made for me because he liked me and because he knew I was the support of my mother. He was that kind. And politics were stronger then than they are today, even. Our wildcat brewery was never knocked off.

This wort place in Waupun was a good-sized stone building. Peterson has been dead a long time now. But he was quite an operator. I knew him when I was a real young kid and my dad was still alive. My dad was sick, he had had a stroke, and during his last years he was showing me how to do everything: how to make the beer and all. And beer was our business. Arnold was a good friend all through that.

And you could make a little money on home brew. The trouble with most of the guys was that they didn't know how to make it. They would get a wild fermentation in a beer, that went skunky, spoiled; buy my dad, being a brewmaster, he knew how to do this. He used the fermented slow process. He had to keep it cool. And there wasn't then any modern refrigeration, so we had "swimmers" and we put ice in them . . . floating refrigerators was what they were, to keep the beer from getting too warm. Beer should be fermented at a low temperature . . . about fifty to fifty-five degrees if possible. Then you will never get a wild fermentation . . . or get it skunky.

Gard: I have heard that there were some Chicago gangsters who moved into the Marsh during Prohibition.

Lechner: Oh yeah. They kind of protected their still houses. This is when they were making alcohol. That was down below Strooks Point. They kept machine guns ready there by their stills all the time. Italian boys mostly. The Chicago gangs didn't move into the Horicon area too strong, and the alcohol business around here was small. There weren't any gang killings or anything like that up here at the Marsh. Us little wildcat operators were too small for the big city boys to bother. One thing that was interesting, the boys from the big city didn't attempt to conceal their distilling operation hardly at all. They just come up here to make alcohol and they did it. You hear about the still back in the southern mountains being so well

Horicon Marsh Family

hidden, and all, but not here in the Marsh. Everybody knew what was going on.

As far as our wildcat brewing business was concerned, about fifty percent of our business was from Milwaukee and out of town. People would drive away out here just to drink our beer, because it was always the same and always good. And we had a real nice little tavern. My dad, as I told you, had been brought up under Kaiser Wilhelm and everything had to be just so. No dust. Glasses always shined and very clean. My dad was so particular that he would look at our fingernails to be sure they were absolutely clean before he would let us go into the tavern. A necktie on all the time, and a clean shirt and collar. We couldn't smoke behind the bar. We couldn't sit on a high chair behind the bar. The cigars in the counter had to be just so. And the bar and the chairs polished.

Gard: Did you serve a free lunch?

Lechner: You betcha. All the time.

Gard: What'd you have?

Lechner: Cheese, the best from all around the country. And we had some fine small cheese factories. It was wonderful. And we had homemade bologna, liver sausage, summer sausage, firefish, called sprats, herring brought in a keg; Limburger cheese and raw beef. Those were the basic things we served at the free lunch. Swiss cheese, brick cheese.

Gard: This lunch. It was absolutely free if you bought something at the bar?

Lechner: Yep. Come in and have a beer and help yourself.

Gard: I always wondered whether it was really profitable to have the lunch.

Lechner: You bet it was. It's funny. When you go through Prohibition . . . it got so we didn't even think that we were breaking any law. We were brought up with beer and liquor. It was a way of life. We didn't change, and we really couldn't accept that we were supposed to.

Ed told me that, though it wasn't exactly cricket perhaps, he performed a real service to the workmen of the Mayville area by opening his tavern at four in the morning, as the shift was just getting off from the die and tool and Maysteel Works. It was the year of Pearl Harbor and everything was going full blast. The workmen had two shifts. Well, while the law technically said that you were supposed to open a tavern at 8:00 A.M., still, it was wartime and the boys needed a little refreshment when they got off work. Ed and his wife put out a fine free lunch, too, and the early opening was so appreciated that they did much more business between four and eight in the morning than they did all the rest of the day.

In that corner tavern in Mayville, Ed averaged five

half barrels of beer a day. There were only three hours out of the day that the tavern was closed . . . only from one in the morning till four in the morning. The rest of the day it was open and going great. Ed would run it from nine until one, and his wife would rise at four and run the tavern till nine. There was good money in punch boards, which were still legal then, and all the boys working in the defense shops were making good money, and they were eager spenders, too.

Once, Ed said, the Ringling Brothers Circus played in Mayville. The Ringling Brothers were real poor in those days. They got their start by going out from Baraboo, Wisconsin, about sixty-five miles from Mayville, with a little wagon show, just a few animals, and they played all the towns all around that part of Wisconsin, and so of course they played in Mayville. Well, a circus was a pretty big event those days, anyhow, when there wasn't so much community entertainment, and the folks were sure glad to see the Ringling boys come to town. Everybody was real proud of the Ringlings, because they had the spunk to start a show and all. So when the Ringlings came to Mayville they were broke. And their animals were hungry. So the merchants got together and took a

collection to buy hay and feed for the animals, and they were able to put on the show. They always said that the Ringling boys felt a big debt to the town of Mayville, and really liked the place. When the show got big, and prosperous, Ringling boys offered to come to Mayville with the whole show, though it was far too small a town, and have a big time, just because of what Mayville did for them when they were poor.

Gard: Would you say, Ed, that the people who live now around the Marsh are any different? Do they behave any differently from other folks you know? I'm trying to get at whether the Marsh has any peculiar character that transfers to the people.

Lechner: I believe that these old Marsh farmers, and their descendents, are real honest people. They are down-to-earth people. They are willing to help you. Say, for instance, a farmer has a fire and the barn burns down. You never see people anywhere that will work as hard day and night to help him get a barn back up. I believe that this warm trait comes from hunting and fishing and trapping and being with nature. I am sure this all has something to do with it. It's hard to put into words. But out here around the Marsh a man gets pretty conscious of fundamental things. The cycles of nature become very plain to him, and he puts his chief faith in things like the seasons, and the changes of color on the Marsh . . . look now, how the willow out there is red, and has a very thin cover of ice on it . . . ever see anything more beautiful than that? Well, a man gets used to beautiful things up in here. He expects to see them, and yet he doesn't take them for granted either . . . and I think that same kind of thing carries over into human relationships. He sees the folks he is familiar with, and he doesn't take them for granted, but maybe he sees them as necessary parts of his scheme of the Marsh and its beauty. So he treats them right. He doesn't want to hurt or destroy them any more than he wants to hurt or destroy the wildlife or the vegetation of the Marsh. You see what I mean? You take the Schabels, or the Wasses or the Voights, or Mrs. Martha Krueger, the Starrs . . . all these people that were born and raised out there on the Marsh. I don't know a one of them that ever was a bad person. Old Eddie Lehner, old Pete Feucht, they are the salt of the earth. They are simply good people. Well, how did they get that way unless it comes from their familiar surroundings? I was talking to old Pete the other day. And he had a fox trap out. And when he went to the trap he had a rabbit in it. And when he got to the trap there was another rabbit sitting outside the trap, waiting for the other rabbit to come out. Old Pete couldn't take it. He had to release the rabbit he had in the trap. This is the way they generally do.

MARSH DANCELAND

Ed ran a tavern and a dance hall business in Kekoskee for thirty years and never called a policeman or a sheriff. And he did some really big business, too. The biggest night he ever had was when the "Silk Umbrella Man" Louis Basoechl and his band from Milwaukee came to play at the dance hall. It was a polka band, of course, and the local folks went big for polka. The dance was sponsored by the Kekoskee Fire Department. Ed had 1,144 paid admissions that night. And that's some dance hall business for a little burg like Kekoskee. Folks couldn't really dance, was so crowded. Just stood and watched the Silk Umbrella Man.

The young folks from the Marsh came into Kekoskee for the dances. Each and every one of the family, and every once in a while Ed would stage a wedding dance for a Marsh couple. The young couples all around the whole Marsh country knew Ed. In fact, he was a kind of godfather to most of them, because through the years he would watch them comin' to the dance hall, and see how they danced and paired up together, and pretty soon he could see that this one or that was getting more and more interested and after a time they would decide to get married.

That was just what Ed had been waiting for. He would, of course, make all the arrangements for the wedding dance. They would ask him if they could rent the hall, and he would always say that they could have the hall free. He would, however, print the posters and see that the posters were sent out, and he would hire the band and sign the contract.

At about nine o'clock the orchestra arrived . . . usually with a lot of whooping and hollering and set up their instruments and had supper. The orchestra always had to be fed first . . . give 'em extra strength because the wedding dances were real active for both orchestra and dancers.

Then, shortly after nine, the wedding party would arrive. There would be a lot of honking and yelling, and really good humor of the most robust kind. When Ed got them inside, and calmed down a little, he lined them up with the bride and groom first and the rest of the party all in couples, and he got the orchestra ready, and they would blow a big fanfare to salute the bride, and the wedding march would start . . . around the hall, maybe two or three times, and then with a yell the procession would break up and the orchestra would bellow into a favorite polka and the dance would be off! And Ed would announce: Everybody dance!

Well, the dancing would go on, hardly ever stopping unless the orchestra just had to quit for a little rest, until about eleven o'clock. Then Ed would announce the grand march. This was when the whole hall of dancers would sort of pair off and march around the hall; and after that was over Ed would announce that there was going to be a dance just for the bride and groom . . . a waltz. They would dance the wedding waltz around the hall, with the bride and groom's folks watching so proud from the sides. And really, the wedding dance was profitable for the young folks because all of the paid admissions went to them as a gift. They would have to pay the orchestra, the ticket sellers, and maybe, if the crowd was real good, they would walk away with two or three hundred dollars. Ed's part of the take would be the sandwiches, the beer, whiskey and the soda water . . . all the bar stuff, plus the wardrobe where they checked their coats and things.

It was all weekend work. Every Saturday you could almost sure count on a wedding dance . . . except during Lent, that is. Nothing ever happened during Lent.

One way that Ed got to be so successful at the dance hall business was that he always tried to hire the polka bands that the young people told him were especially lively. The bride and groom were interested in getting the bands that would pack in the most customers.

Nowadays, though, they do the dances a little different than when Ed was so active. The party will come late in the afternoon now, and they will have a meal right in the dancehall. It isn't advertised for the public. It's for the immediate friends and relatives of the bride and groom. So the dances aren't as big now, as they used to be, but they are still pretty good business around Kekoskee.

Ed told me quite a bit too, about the old iron mine days. Iron Ridge is named because of the iron, of course, but Neda was where the actual mine was.

You would never know now, just driving past the one time busy town of Neda, that it was the center of an iron mining industry. But there was (and perhaps still is) a very considerable iron deposit there. Many men went to the mines to work every day, and, walking back through the fields, up the steep ledge, and back across a corn stubble you see where the earth has caved in over the old mine shafts, and an open airhole more than a hundred feet deep is still open in the middle of the open field. A single piece of rail protrudes from beneath the ledge, now crumbled in, where the old entrance to the mine was. You can't go in there anymore, but it was a good mine once, and there are still old-timers around who worked there. Mayville, not far away, was the center and the blast furnace went full time until the Depression.

Sometimes Signs Protect

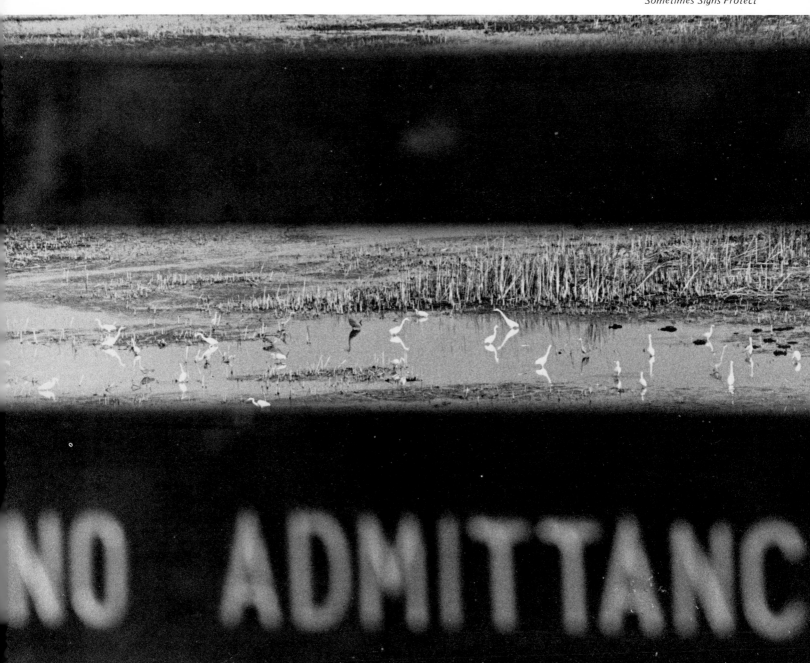

I spent a morning with Bob Personius, manager of the Horicon National Wildlife Refuge at the north end of the Marsh. In a way I was interested to see how his comments might contrast with those made by Jim Bell of the State Headquarters. Bob is personable, articulate, and very, very dedicated to his job. In fact, one of the things that has impressed me most is the diligent attachment that the workers at the Marsh have to their responsibilities. Their enthusiasm is infectious and inspiring in an often too, too blasé attitude toward work in today's world.

Bob Personius said that Canada geese are native only to North America; in fact they live and breed all over North America. Each flock is quite definable in its nesting area, its migration path and its wintering habits. He said that our flock at the Horicon Federal Marsh nest on the western shore of James Bay, and extending part way up the west shore of Hudson Bay. They migrate across the Great Lakes, across Lake Superior and Lake Michigan, and they spend the winter approximately at the confluence of the Ohio and the Mississippi Rivers, mainly in southern Illinois and in nearby Kentucky.

The difficulty has been that these birds have built a tradition to stop at Horicon Marsh in the fall and also in the spring. They are attracted to Horicon because the Marsh has the three primary needs for wildlife, food, sanctuary, and water. They have built this tradition for stopping at Horicon and they couldn't possibly have picked a better place, because their food habits are exactly the same as the holstein cow; we're in the middle of Wisconsin's Dairyland, the goose finds the climate right for the season it has to travel, the food it needs, for the water it needs. The main thing the geese need is food; and there are probably 200,000 acres of corn within ten miles of the Horicon Marsh.

In addition, there are alternate roosting areas for these geese, satellite refuges; several of which have been consciously put there by the State of Wisconsin; but others that are simply fortuitous, the large lakes such as Green Lake and Fox Lake which are excellent roost areas for these geese, and around all these lakes there is corn; so from the goose standpoint it couldn't work out better. But the problem is that the goose is number one on the American waterfowler's want list. He wants to hunt geese, so there is a tremendous demand, and the demand far exceeds the supply. If every hunter who buys a duck stamp and wants to take waterfowl . . . if he was to be given only *one* in a single year, he would decimate the entire flock. You have to control the kill; and this has been one big bone of contention.

The public does not understand this. They don't see why there aren't plenty of geese. Well, the reason is that you can *see* geese. They aren't like rabbits, hiding in the long grass. They are visible. And the hunter or the farmer thinks there must be millions of them. He doesn't understand, then, why we have to limit the killing of geese. But the geese are an international resource which must be shared with other countries, and with other states. The harvest must be carefully regulated. The hunter isn't necessarily happy at this restriction. It's forced on him.

But there is also a threat to the corn crop; and this threat is very real and is getting worse. The farmer in the Horicon Marsh area typically raises corn which he uses to feed his dairy stock. This corn is usually ready for harvest about mid-October; by that time the geese are all here. The Refuge is very small. We don't have sufficient government land to raise enough crops to feed anywhere near the number of geese that stop. The geese are forced to leave the Refuge to feed. They must find farm crops, and normally, it works out all right, because the farmer wants the hunter to keep the goose out of his fields and the hunter wants the geese to come out of the Refuge so they can be hunted. The farmer doesn't mind if the geese will stay in his picked corn. If they glean the picked cornfields that's fine. But if for any reason the farmer can't harvest all his corn right away, if it's wet and he's delayed, then the geese may get into his standing corn, and they can cause serious crop losses.

The goose that habitually comes to the Horicon Marsh is not the largest of the Canada geese. It is probably third or fourth in actual size, but it is still a big goose. The gander should weigh about ten pounds. It can reach up about thirty-eight inches from the ground to strip corn off the stalks. The goose makes no effort to push the stalks down; he just reaches as high as he can with that bill. Very little corn grows out of reach of the goose. He can get most of it.

A goose is supposed to require about one half pound of corn a day. So we now have about ten million goose-use days, that means one goose on any part of the Refuge for one day. So ten million goose-use days in the fall means a requirement of five million pounds of corn. Of course the Canadas also eat many other types of seeds, they eat a lot of green leafy plants like alfalfa, and they eat pasture weeds, and other things.

The potential loss to the farmers is tremendous, and in other parts of North America, and Canada, the

Kids Love Marsh-Time Adventures

losses are extreme. Alberta farmers lose millions of dollars to ducks every year; but here at the Horicon Marsh we have two programs to protect the farmers. One is the actual prevention of damage and in this program, if the farmer will call us, and make us aware that he has a problem, if big flights of geese start actually descending on his corn, we will go out, scare the geese off of his land. To do this we'll leave automatic exploders and scarecrows to assure that the birds do not come back again. And in the event that he actually does suffer loss the State of Wisconsin has a damage claim law. They will compensate the farmer for loss up to $1500 less income from goose blind rental.

So far, losses have been very slight, not more than five to seven thousand dollars a year. But we still have the threat of something serious happening. And the threat is enlarging because of the restriction on hunting. This Horicon portion of this flock is growing at a great rate because it is under harvested.

Bob told me his tale of the great "Goose War" that had reverberations all over the U.S. He told me that in order to understand what lay behind the Goose War, so-called, of 1966, you have to go back to the beginning of the Horicon flock.

There have always been Canada geese in Wisconsin . . . at least since about 10,000 years ago when the Glacier receded. Since about then Canada geese have probably migrated through this area. It's doubtful that they were ever in the abundance that they are

Government Plane Hazing Geese

A Farmer Takes The Problem By The Handle

today. You have to remember that Canada geese are vulnerable to native populations... Indians, Eskimos, and certainly lots of geese were killed. But there were still considerable numbers that migrated through here, and it may be that not many at all stopped at the Horicon Marsh, because this Marsh was a prime hunting and fishing ground for the Indians, and the geese might well have avoided coming.

Also, in more recent time the "market hunters" slew vast numbers of ducks and geese, and that drove the birds off even more. Geese will not stay where they're harassed too much. Apparently, though, this flock did stop in Wisconsin but stayed on the big lakes, probably at Lake Geneva, and then these geese went on down to the Mississippi and Ohio Rivers. They became river geese to start with, and they wintered on the sand bars of the big rivers, and then in the 1920s when the army engineers came up the Mississippi and eliminated the sand bars and channelized everything, the State of Illinois suddenly became aware of the plight of the geese and decided to build refuges; they built the famous Horseshoe Lake Refuge, and the flock responded to the protection and grew much larger.

Then another thing; after World War II when the boys came home, about the first thing they wanted to do was "go huntin'," and they almost wiped the flock out; by the late forties, when the protection programs really got going, it was the start of the great growth. The harvest of the geese was controlled from about that time.

The main mortality on this big bird is man with his gun. Once that bird gets on the wing there is almost nothing else that can hurt him. The fact is, that Mr. Citizen wants geese in his back yard. He wants to hunt geese. He wants the birds around so they will bring tourists. When we got the Horicon Refuge going, in the forties, the geese immediately responded. We had about twenty, thirty, forty thousand, and they wouldn't even budge out of the Refuge. We actually went into the Refuge to hunt them at that time. We put blinds around the edges; then suddenly the program began to grow, and became huge. We all thought what a great thing we had done, because more and more birds were coming. And suddenly it got too big, and that's when the problem started.

We knew then that we couldn't have too many geese at any one Refuge, and we knew that we had to reduce the number of geese at Horicon, and that's how we got into the big hassle in 1966.

As I said before, there are three things that wild animals want. They want sanctuary, water, and food. We had all these things here; and our thought simply was, all right, we'll just remove the food and the sanctuary, and see what happens. In other words, if we took away one or more of his vital factors, would the geese just continue their migration down the Flyway, and so reduce the bird population here?

This was what we wanted. We wanted some of the birds to quit stopping at Horicon. We laid out a program of not having any food on the Refuge, and we made plans to harass the bird, scare, or haze him away. This is a wildlife technique that has worked with geese in other areas.

It was a large and costly undertaking. It takes a lot of men and a lot of material to scare geese. You might not think so, but it really does. We had to have airplanes, amphibious vehicles, a lot of personnel. And then, first, we had to get the agreement of the Flyway Council, because we are in the Mississippi Flyway. And that means that the Flyway Council, which is made up of representatives of eleven states and Canada, would agree that this technique of frightening the geese away from Horicon was desirable. And we did get this agreement. Then we went to the State of Wisconsin. We said: Yes, we're going to do something quite drastic in your state. We want to reduce the number of geese coming to Horicon. The State says, yes, and they agreed to help harass the birds. They were in full agreement that there were too many geese at Horicon for proper management of the flock. And it was all set up.

Until about September 22, 1966, the way was greased. The hazing of the geese was strictly routine, it was a cooperative venture. The bureaucrats had done everything that bureaucrats can do, and suddenly, in the midst of all this, one man, a very important one too, in the State of Wisconsin bureaucracy, said No.

To this day I guess nobody really knows why he did that. There was no question of the need of reducing the goose flock. Everyone had agreed to it. Nobody objected, and it's mighty rare when you can do something as drastic as we were contemplating doing to a wildlife population, and not get one single protest.

Federal Dike Building In Progress

Well, when the State had to decline, then the Federal Bureau decided to go ahead unilaterally and see what we could do on our own. We were ready. We had the materials and the men. We checked back with the Canadian council, and with our Washington office and got the green light.

Our major purpose, our objective, was to deny the Refuge to the birds. The birds were using the Refuge for three things: for a night roost, because the Canada goose roosts at night on water; he was using the Refuge for food and finally, he was using the Refuge for "loafing." He has these three periods in every day. At night he's on the water roosting; in the morning and evening he goes out to feed someplace; and in the middle of the day he loafs. It may be that he loafs in the same place he eats, or it may be back on the water, or it may be in a pasture or a plowed field. So we decided to deny him the use of the Refuge for these three purposes.

First we planted exploding devices here and there in the Refuge, and then with the use of boats, airplanes and just men standing and shooting shotgun blanks in the air, we just hazed, harassed the bird. And the geese responded normally, they left the Refuge, and there wasn't a goose left in the whole Marsh during the daytime.

Now this was before the hunting season; and we had to do it before the hunting season, because what we were doing would simply scare the birds out of the Refuge and over the guns of the hunters who would be waiting on the edges of the Marsh.

Well, the geese moved off of the Refuge, but as soon as night fell, they came back. You can't run the airplanes at night, you can't see geese very well at night. So the program was a total failure. The geese simply moved out to farmers' fields during the day, and it was as though they were saying, Well, I don't need your Refuge anyhow. There's plenty of food in these farmers' cornfields. I'll just come back and roost in the Marsh at night. And it was as if they were really enjoying the game!

It wasn't our fault, either, because the original plan was that we would work only in the Federal Refuge; the State of Wisconsin men would pick the geese up at the border of the Refuge and keep them moving on from there. It was this State of Wisconsin part that wasn't done. We'll never know whether or not this program would have worked with the geese.

So the 1966 story of the year at Horicon Marsh was the goose hazing project. The controversy between the Federal Government and the State developed on September 9, when the Chairman of the Wisconsin Conservation Commission suddenly demanded that the Federal Bureau not go through with the hazing plan. And he was quoted by the press as saying that if we did proceed he would order State wardens to arrest the Federal men, although there had been that agreement between the Wisconsin Department and the Bureau that the hazing could be done. The Commission insisted on legal action when the Bureau decided to go ahead with their hazing program. And on September 26, 1966, the county sheriff stepped into the Federal Refuge Headquarters and handed out summonses and "show cause" orders to a number of our personnel. Whereupon the case was turned over to the U.S. attorney.

Anyone present during September and October when the actual hazing was in progress might have seen twenty-five Bureau employees operating planes, cars, trucks and amphibians, using automatic gas exploders, scarecrows, shell crackers, shotguns and firecrackers. From dawn until far into the night men were prowling over the Marsh, and along its edges, scaring geese. There were, however, enough roosting areas available each night in areas where the vehicles couldn't reach, to accommodate 50,000 geese. Although the hazing temporarily dispersed birds in southeastern Wisconsin, we doubted that it induced many to migrate.

The whole thing had a kind of comic opera aspect because, though we had been summoned, we knew very well that we were never going to go to court. We were warned, of course, that the sheriff was coming, and we all lined up in a kind of chorus line, waiting for him. And the State men were kind of embarrassed about the whole thing. They were all our friends, and suddenly they had to turn into a sort of spy, watching what we were doing, and one State employee was even ordered to follow us around with a movie camera. He was told to get out there and get evidence. And of course, the press descended in droves. NBC called every day from Chicago trying to find out when the best time would be for them to come and film the whole hazing operation. We never heard a word from CBS, but they arrived at just about the same time the NBC crew got here. Never did know how CBS found out about it.

And of course we "demonstrated" our hazing techniques for these television people. We had to go out in boats with the crews. Well, we weren't prepared for them, and all our boats were in use hazing the geese, so I found an old scow and old motor, and took an NBC television crew out on the Marsh, and ran out of gasoline . . . They were kind of frustrated. They were so disgusted they put me into the stern of the boat, and they rowed the boat back to the dock themselves.

You have to understand that man has radically changed the behavior of wildlife. Birds, for example, have simply adapted to things that man has done, 198

that concern their needs. At Horicon we have created large water areas. We have provided the goose with foods that he never found before. This flock feeds to a very large extent on corn. Since the Refuge was established the geese have learned to walk into standing corn. And we have noticed that the geese now will fly right into the cornfield. They wouldn't do that before. The birds would land at the edge of the field and walk in. Now they fly right into the midst of the standing corn.

This is corn that is seven feet tall, with rows that are twenty-eight inches apart. Now the wingspread of the goose is about five or six feet, much greater than the distance between the rows. Yet they fly right in, and fly right back up out of it. You can hear their wings beating against the cornstalks; but this doesn't hurt the bird any.

We have almost 2,000 acres of crop land in the Federal side of the Marsh. The whole Federal holding is 21,000 acres. We plant part of the tillable land to corn and part to winter wheat that we plant in August. That makes a thick, green pasture for the birds to graze on. We have given the local farmers the crop land acreage at the Marsh edges to sharecrop for us. On this land the farmer plants corn and alfalfa. We generally have four to six hundred acres of corn for the geese. The normal pattern is for the geese to come to Horicon in the fall and then to disperse as they follow the food away from the Refuge. They go mostly in a westerly, northerly and northwesterly direction.

One of the worst things that wildlife managers face is the tendency of the public to attribute humanlike qualities to animals and birds. The Canada goose is a very romantic wild creature. It is greatly desired by hunters, of course, but it also has a mystique for people who are not the least interested in hunting. I think it was mostly to do with the mysteriousness of the migration. The goose brings the spring, the first songs and sounds of the coming spring, the harbinger of the end of winter, and the mysterious fact that the wild goose seems to bring with him something of the lure of far places. And then the geese are such creatures of habit. They will return again and again. But the most widely goose-human characteristic is this "pair bond." That's where people really begin to feel related to the goose. Mated for life!

And the goose really does mate for life. But if the pair bond is broken through the loss of the mate, they will definitely select another mate. But it is very true that the gander protects the female, especially while she is on the nest, and both the gander and the goose care for the young. They will also care for the young of other geese, in so-called gang broods. In other words, the geese have characteristics that people approve of. By comparison, the male ring-necked pheasant has a big harem. A mallard duck only has one mate, but just as soon as the ducklings are hatched, he's gone.

One of the main things we do now is to display our great goose show to the public. We have in the Marsh one of the great wildlife shows in our nation. Every fall especially, great numbers of people come, in 1971 over 300,000 in October alone. They line Highway 49 for miles. It's so congested now that it's grown very dangerous to come to watch the geese. All this takes place along a sixty-five-mile-an-hour highway. So far no one has been hurt. Now, however, we are planning a new road, actually using old abandoned trails, down through the Marsh, so that people can drive out to where the birds actually are.

It's a national policy not to allow snowmobiles on the Federal Refuge. We believe that during the winter, wildlife need solitude. I hope we can keep it for them.

Geese, of course, are not the only birds that use the Great Marsh for food and refuge. There are egrets at the Marsh. Harold Mathiak, research biologist at the State Marsh Headquarters told me that people coming to visit the Marsh are always excited about the egrets. He figures that one egret might cause about as much interest as fifty ducks. He says that you can just stand and watch egrets and they don't appear to be doing anything, then all at once one will spear down and come up with a fish. They don't seem to be watching for fish at all. And Harold says he has

never seen an egret catch a bullhead. The bullhead is a spiney fish and could be hard to swallow. But Harold has seen blue herons catch bullheads lots of times, and they can apparently not only swallow the fish but can regurgitate it when they reach the nest and have to feed the young herons. Herons can swallow bullheads, but he has never seen an egret do it.

There are about 300 egrets breeding at the Marsh in an ordinary year . . . not so many, but there are only two places in Wisconsin where the egrets breed, at least that Harold knows about. One place is the Horicon Marsh. The other is the egret colony at Fox Lake. There are, however, several egret colonies on the Mississippi River . . . at the edge of Wisconsin.

Harold said that when he first came to live on the Marsh he lived on the old Mieske farm, over on the west side. That was the farm that was owned by the old man who harvested all the ducks for commercial use in the old, old days (before 1900). And when he was living out there, Harold said, the Marsh was so wild and beautiful. He was outside all the time. There would occasionally be great flights of pintails that would be leaving the Marsh about dark, and then would return a little later. The flights would pass directly over Harold's house, and the whole family would be out watching the spectacular sight. And perhaps twenty years ago there would be thousands of redheads, massed often, on a very small expanse of water. Unfortunately the redheads are mostly gone now.

But Harold said that he thought the most interesting bird on the Marsh was perhaps the yellowhead blackbird. We are mostly on the eastern range of that strange and very persistent creature. Once you have heard a yellowhead, said Harold, you will never forget it. To me it is like a crazy sound from the kind of absurd world we are all supposed to live in . . . it is a song that has no logic, no beginning and no end. It is just there but entirely wild and beautiful. Harold says that most visitors want to see and hear a yellowhead and he always tries to see that they get their wish!

Sportsmen's clubs of the Horicon area are especially concerned with the wildlife on the Marsh. Many pheasants are raised each year by the clubs and released, and the club members bring to life the great sports days of the Diana Club in their regard for the wildlife and for the Marsh, and the training of excellent bird dogs.

On the Clark Farm Waterfowl Area, where many interesting experiments are carried on in association with sportsmen's organizations, the clubs actually raise funds to aid in the propagation of wildlife. Special nesting places for mallard ducks provide safety and more comfortable nests for duck mothers.

Gene Mauch, President of Wetlands for Wildlife, which had its beginnings at the Horicon Marsh says that observing the thousands of wildfowl, and the exciting variety of bird life at the Horicon Marsh Wildlife Refuge inspired a group of men to arrange for the incorporation of the nationally recognized non-profit organization of Wetlands For Wildlife with present headquarters at Mayville, Wisconsin, which is dedicated to the promotion, acquisition and maintainance of all wildlife habitat. In this original group were Ben Boalt, presently President Emeritus; Director Owen J. Gromme, retired Curator of the Milwaukee Public Museum and national recognized wildfowl and bird artist; Lester Dundas, former manager; and James Bell, Manager of the Federal and State areas of the Horicon Marsh; Arthur Molstad and William Frankfurth of Milwaukee and Richard C. Bonner of Grafton, Wisconsin.

There are least bitterns in the Marsh, too, but you have to be really careful or you will scare them. They are extremely timid birds. Not so many folks have seen a ruddy duck up close and this duck is about the most dressed up feathered thing alive, Harold thinks. The bill of a male ruddy duck in breeding plumage just doesn't look real, that's all.

But in breeding season birds are quite tame as a rule, and strangely enough, traffic doesn't necessarily drive birds away. If birds get used to people coming to view them, well, the birds just sort of ignore the people. The trick is, when you are walking through the Marsh, and that's the best way to see things happen, to just listen all the time. You will hear so, so many wondrous things! And see things, too. On the shores of Four-Mile Island for instance, you may still see the bank lines of old Lake Horicon! That should make anyone stop and think and remember. And that island is where the heron rookery is located. Harold counted 976 nests of the blue heron, egret and the black crowned night heron there in 1971.

"Four-Mile Island" is an historic name on the map of the Marsh. Just who first called it "Four-Mile" is probably not known by any living person. It does, however, have another, and more romantic name, "Little Everglades of the North."

The island is absolutely off limits during the bird nesting season. Nobody is allowed to walk on the island, but from a canoe, or by using an occasional tour from one of the boat tour companies at Horicon, a good view of what goes on on old Four-Mile can be had. The island is where most of the herons, egrets, the long-legged waders can be seen; and around the island there is plenty of other wildlife. Mink and even

otter can sometimes be viewed in the area and beaver, racoon and an occasional fawn, for many fawns are born each year on the little Everglades.

Speaking of hiking, for anybody who really likes to walk, and as Harold says, it is the best way to really see what goes on at the Marsh, there are several hiking trails that can be followed.

One of these, called One-Mile Island Trail starts at Arndt's Ditch parking lot, goes north across One-Mile Island to the main river channel, then turns northwest along a dredged ditch to Kohlman's Fence, and then returns.

At the State Marsh Headquarters on Palmatory Street, you will find signs showing you where you may hike. There may well, however, be restrictions on certain breeding grounds. You should inquire, if the headquarters building is open, about these closed areas.

You can see a lot of wildlife on the Indermuehle Island Trail. Off Highway 28 turn north across Indermuehle Island . . . and then there is the Federal Dike Road. You turn west at the Federal Dike sign on highway TW. You drive as far as you can, then you have to get out and walk. That's the best part any-

204

way. And there are man-made impoundments that Marsh staff will direct you to. Wildlife abounds along these impoundments.

And frogs! They don't have any bullfrogs now at the Marsh. Harold thinks the reason is that there are so many herons and that the herons simply eat up the bullfrog population. Both bullfrogs and green frogs are rather slow-moving creatures, and perhaps somewhat stupid compared with the smaller frogs. Harold has, many times, walked right up to a green frog and touched it. But a frog that lets a heron touch it is a dinner for that heron. That's all. Harold has often tried to get green and bull frogs started at the Marsh without much result. Too many herons. But there are plenty of leopard frogs . . . they have a real plague of leopard frogs at the Marsh in some years. A leopard is about four inches long when he's sitting, and many years ago there were so many people who came to the

Marsh just to go frogging. The frogs seem to flow with the river and end up in the deep holes. The hunters would follow the river in winter and chop holes in the ice here and there, where the deeper holes were, and they would have a rake with some hardware cloth on it, and if they were lucky and found a frog hole they could just rake the frogs out. They might sometimes get a hundred pounds at a time, and they sold the frogs to biological houses. You have to chop a lot of holes in the ice to do it that way, and it takes a lot of testing to be sure where the frogs are!

Young people in general love the Marsh, and once in a while a younger fisherman will hook into a beauty. There are trout in the Marsh streams still, and now with the new program maybe they will come back in numbers that might make even the old Indians envious.

Fred Lechner, Kekoskee With 22-Inch 3¾ Pound German Brown He Caught

Kids especially love the young owls. Once in a while a lucky youngster gets a chance to feed them, or a young fellow succeeds in making a devoted friend of one of the bolder crows, and rowing or just drifting through the ditches and along the river in the summer stillness is part of everybody's keen pleasure.

For the interest of bird watchers who may come abirdwatchin' to the Great Horicon Marsh, I include the latest list from the Federal Marsh Headquarters of all the wild birds that the crews up there have spotted lately.

Birdwatching is encouraged. This list contains 198 birds which you can expect to see in season. Another 41 birds have been reported but are not normally expected to be present. Season and relative abundance are coded as follows:

PR — permanent resident (present year round)
SR — summer resident (present in spring, summer and fall)
WR — winter resident (present in fall, winter and spring)
TV — transient visitant (present in spring and fall during migration)
WV — winter visitant (present in winter irregularly)

a — abundant (should always see in large numbers)
c — common (should always see but not in large numbers)
u — uncommon (always present but not always seen)

Pied-billed Grebe	SR-c	Bald Eagle	TV-u
Double-crested Cormorant	SR-u	Marsh Hawk	PR-c
Great Blue Heron	SR-c	Sparrow Hawk	TV-c
Green Heron	SR-u	Ring-necked Pheasant	PR-c
Common Egret	SR-c	Gray Partridge	PR-u
Cattle Egret	SR-u	Sandhill Crane	TV-u
Black-crowned Night Heron	SR-c	King Rail	SR-u
Least Bittern	SR-u	Virginia Rail	SR-u
American Bittern	SR-c	Sora Rail	SR-c
Whistling Swan	TV-u	Common Gallinule	SR-c
Canada Goose	TV-a	American Coot	SR-a
Snow Goose	TV-c	Semipalmated Plover	TV-u
Blue Goose	TV-c	Killdeer	SR-c
Mallard	SR-c	American Golden Plover	TV-u
Black Duck	SR-u	Black-billed Plover	TV-u
Gadwall	SR-u	American Woodcock	SR-u
Pintail	SR-u	Common Snipe	SR-c
Green-winged Teal	SR-u	Upland Plover	SR-u
Blue-winged Teal	SR-c	Spotted Sandpiper	SR-u
American Widgeon	TV-c	Solitary Sandpiper	TV-u
Shoveler	SR-u	Greater Yellowlegs	TV-c
Wood Duck	SR-c	Lesser Yellowlegs	TV-c
Redhead	SR-c	Pectoral Sandpiper	TV-u
Ring-necked Duck	TV-c	Least Sandpiper	TV-u
Canvasback	TV-u	Dunlin	TV-c
Lesser Scaup	TV-c	Long-billed Dowitcher	TV-u
Common Goldeneye	TV-u	Stilt Sandpiper	TV-u
Bufflehead	TV-u	Semipalmated Sandpiper	TV-u
Ruddy Duck	SR-c	Wilson's Phalarope	SR-u
Hooded Merganser	TV-u	Herring Gull	TV-c
Common Merganser	TV-u	Ring-billed Gull	TV-c
Sharp-shinned Hawk	TV-u	Forster's Tern	SR-c
Cooper's Hawk	TV-u	Black Tern	SR-c
Red-tailed Hawk	PR-c	Mourning Dove	PR-c
Red-shouldered Hawk	TV-u	Yellow-billed Cuckoo	SR-u
Rough-legged Hawk	WR-c	Black-billed Cuckoo	SR-u
Golden Eagle	TV-u	Screech Owl	PR-u

Great Horned Owl	PR-u	Warbling Vireo	SR-c
Snowy Owl	TV-u	Black-and-white Warbler	SR-c
Short-eared Owl	PR-u	Prothonotary Warbler	TV-u
Common Nighthawk	SR-u	Golden-winged Warbler	SR-u
Chimney Swift	SR-u	Blue-winged Warbler	SR-u
Ruby-throated Hummingbird	SR-u	Tennessee Warbler	TV-c
Belted Kingfisher	SR-u	Orange-crowned Warbler	TV-u
Yellow-shafted Flicker	SR-c	Nashville Warbler	SR-u
Red-headed Woodpecker	SR-u	Parula Warbler	TV-u
Yellow-bellied Sapsucker	SR-u	Yellow Warbler	SR-c
Hairy Woodpecker	PR-u	Magnolia Warbler	TV-u
Downy Woodpecker	PR-c	Cape May Warbler	TV-u
Eastern Kingbird	SR-c	Black-throated Blue Warbler	TV-u
Great Crested Flycatcher	SR-c	Myrtle Warbler	TV-c
Eastern Phoebe	SR-c	Black-throated Green Warbler	TV-c
Least Flycatcher	TV-u	Cerulean Warbler	SR-u
Traill's Flycatcher	SR-c	Blackburnian Warbler	TV-u
Eastern Wood Peewee	SR-c	Chestnut-sided Warbler	SR-u
Olive-sided Flycatcher	TV-u	Bay-breasted Warbler	TV-u
Horned Lark	PR-c	Blackpoll Warbler	TV-u
Tree Swallow	SR-a	Palm Warbler	TV-c
Bank Swallow	SR-c	Ovenbird	SR-c
Rough-winged Swallow	SR-u	Louisiana Waterthrush	TV-u
Barn Swallow	SR-c	Northern Waterthrush	TV-u
Cliff Swallow	SR-c	Mourning Warbler	TV-u
Purple Martin	SR-c	Yellowthroat	SR-c
Blue Jay	PR-c	Wilson's Warbler	TV-u
Common Crow	PR-c	Canada Warbler	TV-u
Black-capped Chickadee	PR-c	American Redstart	SR-c
Tufted Titmouse	PR-u	House Sparrow	PR-a
White-breasted Nuthatch	PR-c	Bobolink	SR-c
Red-breasted Nuthatch	TV-u	Eastern Meadowlark	SR-c
Brown Creeper	TV-u	Western Meadowlark	SR-c
House Wren	SR-c	Yellow-headed Blackbird	SR-c
Long-billed Marsh Wren	SR-c	Redwinged Blackbird	SR-a
Short-billed Marsh Wren	SR-c	Baltimore Oriole	SR-u
Catbird	SR-c	Rusty Blackbird	TV-c
Brown Thrasher	SR-c	Brewer's Blackbird	SR-u
Robin	SR-a	Common Grackle	SR-a
Wood Thrush	SR-c	Brown-headed Cowbird	SR-c
Hermit Thrush	TV-u	Scarlet Tanager	SR-u
Swainson's Thrush	TV-u	Cardinal	PR-u
Gray-cheeked Thrush	TV-u	Rose-breasted Grosbeak	TV-c
Veery	SR-u	Indigo Bunting	SR-c
Eastern Bluebird	SR-u	Dickcissel	SR-u
Blue-gray Gnatcatcher	SR-u	Evening Grosbeak	WV-u
Golden-crowned Kinglet	TV-u	Purple Finch	WV-u
Ruby-crowned Kinglet	TV-u	Common Redpoll	WV-u
Cedar Waxwing	PR-u	American Goldfinch	PR-c
Northern Shrike	WV-u	Rufous-sided Towhee	SR-c
Starling	PR-c	Savannah Sparrow	SR-c
Yellow-throated Vireo	SR-u	Grasshopper Sparrow	SR-u
Solitary Vireo	TV-u	Hanslow's Sparrow	SR-u
Red-eyed Vireo	SR-c	Vesper Sparrow	SR-c
Philadelphia Vireo	TV-u	Lark Sparrow	SR-u

Slate-colored Junco	WR-c
Tree Sparrow	TV-c
Chipping Sparrow	SR-c
Clay-colored Sparrow	TV-u
Field Sparrow	SR-u
Harris' Sparrow	TV-u
White-crowned Sparrow	TV-u
White-throated Sparrow	TV-c
Fox Sparrow	TV-u
Lincoln's Sparrow	TV-u
Swamp Sparrow	SR-c
Song Sparrow	SR-c
Lapland Longspur	WR-u
Snow Bunting	WR-u

Birds which have been seen at Horicon but are either no longer present, are not normally found in this area, or do not normally stop here on migration.

Common Loon	Yellow-breasted Chat
Horned Grebe	Buff-breasted Sandpiper
Red-necked Grebe	White-rumped Sandpiper
White Pelican	Black-necked Stilt
Little Blue Heron	Marbled Godwit
Glossy Ibis	Hudsonian Godwit
Snowy Egret	Northern Phalarope
Yellow-crowned Night Heron	Bonaparte's Gull
White-fronted Goose	Common Tern
Brant	Barn Owl
Oldsquaw	Barred Owl
White-winged Scoter	Long-eared Owl
Red-breasted Merganser	Saw-whet Owl
Goshawk	Whip-poor-will
Broad-winged Hawk	Red-bellied Woodpecker
Osprey	Yellow-bellied Flycatcher
Peregrine Falcon	Boreal Chickadee
Pigeon Hawk	Winter Wren
Bobwhite	Carolina Wren
Ruddy Turnstone	Mockingbird
	Loggerhead Shrike

The marker branches on the oak at the Marsh's edge do have something to point to, now. Perhaps because the Marsh is unique in that it has had greater fertility, and less pollution than other marshes, it was able to recover from the terrible ordeals man has put it through in his foolish attempts to dam, to exploit, to grow crops, to operate hunting clubs, to drain and finally, to restore. Now, man, for once, has been doing the right things, cautiously and well.

The Marsh is an immense recycling operation that can only work if left to itself. The ditches on the Marsh have closed ends and retain the water during the growing season, but within the waters of the Marsh the growth of plants and animals progress on and on. The perennials die and are recycled: the relationships of predators to victims is in balance: a goose is wounded by a hunter's shotgun and escapes into the tall grass. A fox finds the disabled goose and kills it. Trappers harvest and control the muskrats, and the muskrats control the cattails making openings valuable for ducks. The trappers prevent muskrat death by overpopulation. Snapping turtles sometimes seize the floating coots from below. The owl talons the baby raccoon. Pike seize the ducks and frogs. Hawks swoop on rabbits and small birds. The wild geese loose tons of manure into the waters, yet the waters apparently process the defecation of the birds. Everything is living, dying, reproducing. The Marsh is working, everything for the moment seems to be in balance.

Now, at last, the great carp control program is going into gear. Next year, the years after, fine fishing may be ordinary again; and fishermen will hurry to the Marsh. The red man, however, will never return. Yet despite the ironical absence of our red brothers whose storehouse the Marsh always was until the white man took it from him ... despite the fact that there is no present Marsh Indian to relate to the

effigy mounds, perhaps the Great Marsh has still come into its own as a new kind of place for people. Perhaps finally the white man has learned to live in harmony with the Marsh as did the Indians. If the wild goose is the symbol of the freedom of nature, then probably it is fitting that a public so hungry for the wild sights and sounds of nature can find these things in the Marsh. It is rich in inspiration and beauty, and the city man, and indeed, the rural man often living in the midst of nets of concrete highways around his farmland, has little enough left of primitive beauty. If the Marsh becomes a singing harbinger of spring and puts forth the signs of fall in a particularly beautiful way, enhanced by man, then it has served its purpose for people. All of the effort that State and Federal Government has put forth to preserve and restore the Marsh may have been rewarded by just this thing: that man has been made a little bit more aware, more sensitive, and more responsive to earth and sky than he might have been had the Marsh not existed.

Somehow as I stand by the side of the Marsh of a spring evening, I have a dream vision of a long, long line of people beginning with the earliest primitives and Copper People and their successors the Woodland Mound builders and stretching up through and across the Marsh, clear from the post glacial times when the first men arrived to fish in the cold blue waters, to the curious and awestruck crowds gathered today in cars and on foot along Highway 49 east of Waupun, along the north margin of the Marsh, to stare at the Canadas resting in the meadows. And I seem to hear ancient drums; see a group of naked Effigy Mound builders at work creating with earth the relief shape of a sacred effigy. And the drums go on and on, through ages and ages, through stone and copper and iron . . . through primitive drawings on a limestone wall, through bone and spear and flint arrow.

We will be back, the old woman had said when she and her people were led from the Marsh in 1850. We will be back, for the Indian peoples cannot forget our Marsh homeland. We will be back to meet again the wild geese in the spring; and we will be back to follow our ancient trails and our lodges will stand again where they have been our villages against the sides of the Marsh, and the trees will mark our trails.

Oh, ancient Marshland redmen, oh, forgotten ones! I think in my dream your story and tragedy is what moves the heart when alone one stands in the Marsh and remembers the quality of its strange wilderness. It is you whose footsteps I hear receding away into time, fainter and more faint until they are lost, lost eternally. And I have searched for one of you who could tell me what now lies in the heart of a great-great grandson or great-great granddaughter of those old ones whose lands the Great Marsh once included. And often, in my dreaming, and unable to discover even one who could so speak, I have attempted words which you might have uttered, or might utter now. Backgrounding the inadequacy of my language is the moaning of the women and the crying of children, and the wordless pain of the men when they were told to depart for alien lands, more than 100 years ago. And in the background of my imagined words I hear the weeping of the Indian women in joy when they had slipped away from alien fields to which they had been sent in the north and west and had returned over slow, long and bitter trails to the Marshland home in the 1880s, to find all changed, of course, and their joyful tears suddenly blotted with shock as they stared at the white hunters of the old Diana Club waiting in blinds in their wild rice.

Oh, white man, I hear you cry in essence, now, as I dream, leave to us something of the old way . . . of the sky-filled bodies of the birds, of the odor of the wild Marsh, of the spirit of us, the ancient peoples, and of our trails which led where we must go, here and there among our temples and our foodlands. For we too felt and dreamed, oh brothers, and our trails were good and our markers straight. Help our spirits to remain at the Marsh forever, a part of what you have now made yours, but which is still, nevertheless ours also, and every man's. Leave us the movement of the birds and the sound of the wind in the bending grass. . .

And around all, and always, there in a thousand shapes and aspects of evening and morning; among wild grasses and remembered towering wild rice, pervades a kind of Marsh spirit that is absorbed and made a part of me as I dream and watch. The small evening birds flutter across the sky and a thin mist rises from a far part of the Marsh. The sound of drums rises a little and then fades far, far away. Curiously, the drum sound rumbles almost the same low thunder as a far, far jet plane. I wait. Watch. The Marsh, too, fades and becomes a part of the night. I am alone. But in a last tiny flash of vision I see again the twisted branches on the marker oak, and I know that I, through what I have revealed of the Marsh and its story, and through a criss-crossed, restless pattern of sky-beating wings, have seen at least within my own imagination, a moment of what the marker branch has meant.

Marsh At Twilight

PHOTOGRAPHER REFLECTS

If you've never been on the Great Marsh in the evening, Ed said, you may be missing life's finest hour.

And if you have, you may find it difficult to describe that experience.

Inspired by the call of Spring, Dorothy and I trekked out to Townline ditch near the Ernie Bleifuss farms shortly after sundown to see and hear the sights and sounds of budding nature.

Horicon Marsh is probably the best place in America to do this.

An almost complete absence of breeze permitted greening willows to reflect in quiet waters like a huge image in a man-made mirror. Unlike the imperfections in man-made objects, this was nature-perfect, complete with sound effects.

A symphony amplified by Marsh creatures. Millions of crickets, frogs, egrets . . . thousands of small birds, many varieties chirping lustily, and mallards and teal drifting serenely.

In the distance a V takes shape on the quiet water temporarily distorting the willowy images reflected by the clear sky overhead. Still enough light for eyes now adjusting to night vision, but hardly sufficient for available light photography.

Not more than 30 feet before us emerged the maker of that V. I'll call this muskrat Tom. Tom entertained us for a quarter hour as he silently worked over one clump of sod after another along the fenced shoreline.

Then, as quietly as he had come, Tom disappeared.

There were no dull moments.

Overhead a quintet of Canada geese announced their arrival with loud honking, while a loon in the far distance uttered his wild cry.

Seconds later the geese set their wings for a perfect landing on the mirrored surface.

Just like in the time of creation, commented Dorothy so softly I hardly heard what she said. I was too busy just feeling.

—EGM

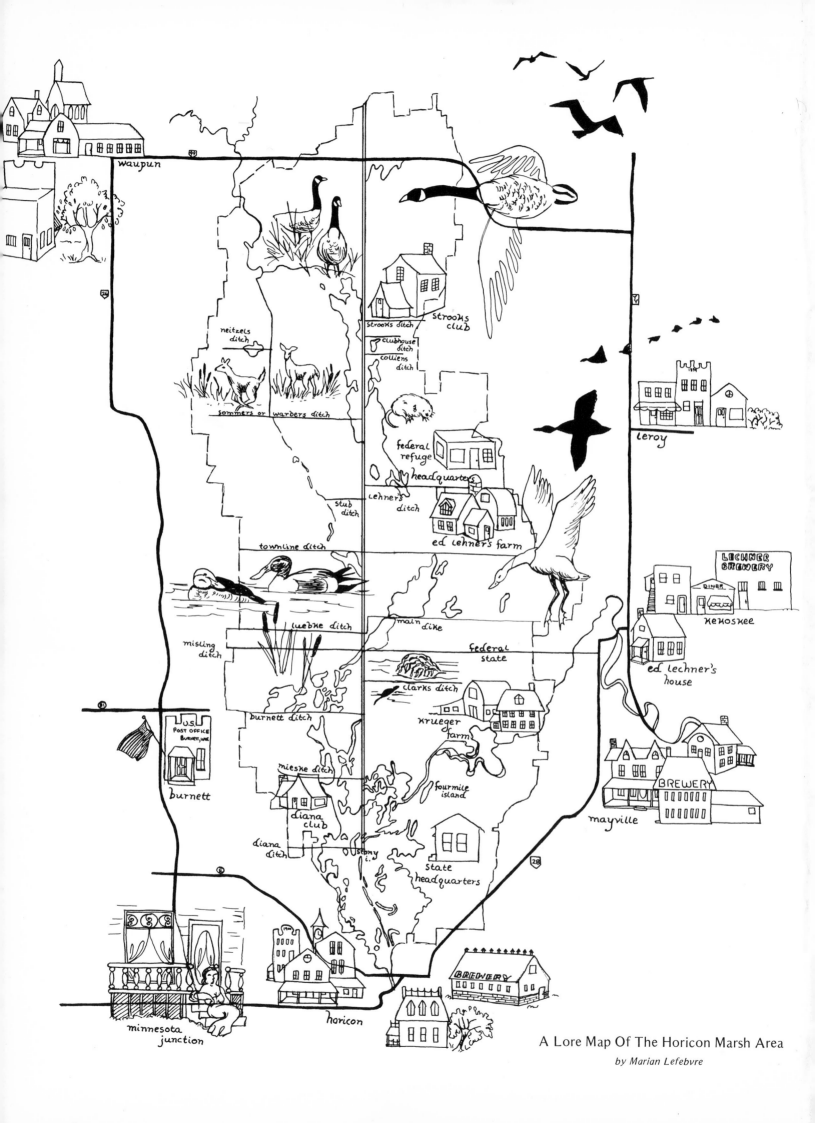

waupun

neitzels
ditch

sommers or warbers ditch

strooks ditch

strooks
club

clubhouse
ditch

colliens
ditch

federal
refuge

headquarters

lehner's
ditch

ed lehner's farm

stub
ditch

townline ditch

luebke ditch

main dike

federal
state

misling
ditch

clarks ditch

burnett ditch

U.S.
POST OFFICE
BURNETT, WIS.

krueger
farm

burnett

mieske ditch

fourmile
island

diana club

diana ditch

stony
i.

state
headquarters

minnesota
junction

horicon

leroy

LECHNER
BREWERY

DINER

kekoskee

ed lechner's
house

BREWERY

mayville

BREWERY

A Lore Map Of The Horicon Marsh Area

by Marian Lefebvre